Probing The Limits

Probing the Limits

Collected Works on the Second Law of Thermodynamics and Special Relativity

Germano D'Abramo

First edition: January, 2017
165 pages, 5.5×8.5 inches
This book has been typeset with LaTeX

Cover image by the author.

The Secret is in the Light

Contents

List of Figures

Acknowledgements

My never-ending gratitude goes to lifelong friends Gianluca Capriotti and Gianpietro Summa, the sharpest critical minds I have ever met. The support and encouragement of Prof. Daniel Sheehan (University of San Diego, USA) through the last eight years are heartily acknowledged.

Preface

[...] it is demonstrated that our reason can absolutely not find the truth save by doubting, that it distances itself from truth whenever it judges with certainty, and that not only does doubt serve to uncover the truth (according to Descartes's principle [...]), but that truth essentially consists in doubt, and whoever doubts knows, and knows as much as one can know.

Giacomo Leopardi, Zibaldone (8 Sept. 1821)

I basically do not subscribe to the proverbial saying that "rules are made to be broken". Nevertheless, when such "rules" are limits imposed by physical theories, I strongly contend that they should be probed relentlessly: every limit, even those that appear to be more solid than diamond. Moral obligation to critical thinking should urge every scientist not to trust anything completely, and not to be satisfied with what someone else has told him or her, no matter how authoritative. Physical theories are, after all, the product of human beings; they are human models through which we try to describe in the simplest possible way (to us) how Nature behaves. They are surely not about *how Nature really is*, provided that talking about how Nature *really is* makes any sense at all. Every product of human beings, even the most intellectually sophisticated ones, is always limited and tentative. Moreover, in taking the above critical stance, every genuine scientist should not feel afraid or shamed, because nobody holds the absolute truth. If a scholar is intelligent and lucky enough, his or her own new physical theory can only dream of being the most durable *human model* of Nature's "behavior", nothing more[1].

---

[1] In fact, any intellectually honest scientist knows with 100 percent confi-

It is exactly with this feeling and motivation that I pursued the scientific research that is described in this book. The book gathers together the chief results of the research work that I have carried out on the second law of thermodynamics and the theory of special relativity since 2008. The main part of this book, the first six chapters, is devoted to my research on the epistemological status of the second law of thermodynamics and the connection between thermionic/photoelectric phenomena and the second law. I hope to have provided evidence that thermionic emission could in principle violate the second law: more precisely, the photoelectric emission triggered by the high-frequency tail of black-body radiation at (uniform) room temperature can be harnessed to charge up a capacitor and provide readily usable energy from a single heat reservoir.

Almost all of what is written in these first six chapters is actually the content, updated and harmonized as much as possible, of several research papers[2] that I have published during these years in the following peer-reviewed international journals: *Physics Letters A* [126], *Physica A* [127, 128], *Foundations of Physics* [134], *Studies in History and Philosophy of Modern Physics* [135] and *Entropy* [137]. I must anticipate that in some chapters, I found myself in need of repeating the same concept and/or mathematical preliminary in order to make the reading, according to my judgment and personal taste, as self-contained as possible, and thus more intelligible. Obviously, there are two sides to every question. The downside is that the careful reader might sometimes have the impression of *déjà vu*. I promise that it will not happen too often.

Chapter 7, instead, contains my reflections on special relativity.

---

dence that *every* physical theory, present and future, can never be a complete and coherent description of reality, i.e. it can never be the *truth*: this is in the nature of scientific knowledge. Moreover, this is the reason that makes me laugh (a bitter laugh, though) when I come across some gurus of science outreach (both scientists and journalists) or, more often, when I come across philosophically poorly endowed partisans of science (eleventh-hour positivists) who, spouting self-confidence and unseemly enthusiasm, talk to people as though they, and science in general, had the truth in their pockets.

[2]Where it applies, the material has been reprinted with the permission of the original publisher and proper credit is given in the text.

It is surely the most speculative part of the book (*caveat lector!*) and has never been published elsewhere. I describe two thought experiments on time dilation in the framework of special relativity: the first, about unknown, involves two satellites that orbit about a planet at high speed on nearly equal circular orbits and in opposite directions. I shall show that this scenario provides a perfectly symmetrical version of the twin paradox, and I critically evaluate the possibility of any resolution on the basis of fundamental physics principles, like the equivalence principle, and also by considering GPS satellite data. In the second thought experiment, I linger on the issue of "reality" of Lorentz transformations, and thus on the "reality" of time dilation in Einstein's theory of relativity. Incidentally, in the context of special relativity, I also try to answer an apparently unrelated question: when ascribed to light, can the term "velocity" have the same meaning we intend for ordinary material bodies? I shall show that the attempt to answer this question brings with it interesting consequences. The main contention in this part of the book is that if both postulates of special relativity are assumed to hold concurrently, then the prediction of asymmetric ageing made by Einstein in his 1905 relativity paper appears to be in fact incompatible with them; and the fact that time dilation (which is intimately related to "asymmetric ageing") seems to have been experimentally confirmed provides, paradoxically, a refutation rather than a confirmation of the theory of special relativity, at least as is interpreted today. According to our analysis of the logic behind special relativity, no time dilation should occur under any circumstances. This points to a twofold contradiction: time dilation seems to have been actually observed in several experiments (and should not; first contradiction), but only in some and not in all the circumstances allegedly predicted by the theory (if the theory were actually correct, time dilation should occur in all the circumstances predicted by the theory itself; second contradiction). Finally, in the first of the appendices to this book, a critical assessment of Purcell's basic explanation of magnetic forces, which basically relies on special relativity, is also given.

I strongly doubt that what I have written in Chapter 7 will be subscribed to by other scholars and that one day or other it will receive experimental support: unfortunately, it does not yet have the shape of a quantitative theory that can be tested experimentally. On the contrary, I have high expectations that what I have presented in the first six chapters will be verified or disproved one day by laboratory experiments that are, in my humble opinion, already within reach.

Germano D'Abramo
Albano Laziale, Fermo
December, 2016

Chapter 1

Introduction

A theory is the more impressive the greater the simplicity of its premises, the more different kinds of things it relates, and the more extended its area of applicability. Therefore the deep impression that classical thermodynamics made upon me. It is the only physical theory of universal content which I am convinced will never be overthrown, within the framework of applicability of its basic concepts.

Albert Einstein

In its classical and phenomenological formulation, the second law of thermodynamics states that "it is impossible to construct a device that, operating in a cycle, will produce no effect other than the extraction of the heat from a cooler to a warmer body" (Clausius formulation) or, equivalently, that "it is impossible to construct a device that, operating in a cycle, will produce no effect than the extraction of heat from a reservoir and the performance of an equivalent amount of work" (Kelvin-Planck formulation).

A more quantitative statement of the second law could be formulated only after the introduction of the thermodynamic state-function entropy S (Clausius, early 1860s): it is defined up to an additive constant as the following integral through a quasi-static reversible path (Fig. 1.1):

$$\Delta S = S_B - S_A = \int_A^B \frac{\delta Q}{T}, \tag{1.1}$$

where A and B are two equilibrium states of a thermodynamic

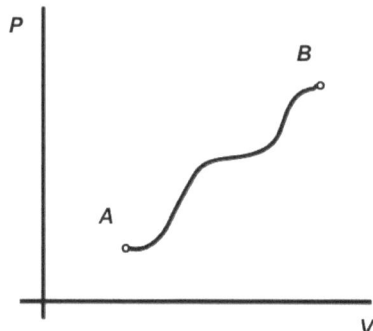

Figure 1.1: Transformation of a thermodynamic system from state A to state B through a reversible path in the pressure-volume PV diagram. The third thermodynamic variable, temperature T, depends upon V and P through the equation of state of the system (Reprinted from [135], with permission from Elsevier)

system, T is the absolute temperature of the system and δQ denotes an incremental reversible transfer of heat into that system (if heat is transferred out, the sign would be negative). The second law could then be stated in the well-known *increasing entropy* formulation: whenever an adiabatically isolated system evolves from equilibrium state A to equilibrium state B, the variation of entropy ΔS cannot be negative; $\Delta S \geq 0$ [19].

From a historical perspective, the second law of thermodynamics originated in the context of classical thermodynamics. Classical thermodynamics was essentially born in the nineteenth century, and its original goal was to understand the fundamental principles that underlie the operation of heat engines with the aim of making them as effective as possible. Note that the 1824 seminal work[1] by Sadi Carnot, who is considered as the father of thermodynamics, is almost exclusively on heat engines and engine efficiency. Heat engines are machines having almost always a gas as working substance and performing cyclic transformations that can be represented as

---

[1]Carnot, Sadi (1824). *Réflexions sur la puissance motrice du feu et sur les machines propres à développer cette puissance.*

closed loops on the pressure-volume (PV) diagram. Therefore, the main subject of classical thermodynamics is the study of heat and work exchange between the environment and systems whose state is uniquely defined by volume, pressure and temperature, together with the equation of the state of the working substance. Moreover, one cannot deny that early classical thermodynamics mainly dealt with a simplified type of thermodynamic system, a gas inside a cylinder performing V, P, T transformations. In that context, the second law of thermodynamics (Kelvin-Planck formulation) appears to be a direct and inevitable consequence. Let us explain. We all know that a transformation must be cyclical in order to be practically exploitable as a source of work. Furthermore, the amount of work performed by a gas[2] in a cycle corresponds to the area inside the closed loop on the PV diagram that represents the cyclic process—the work is positive, i.e. released, if the cycle is performed clockwise. Thus, the only possibility for a cyclic process to perform work is to be a closed loop enclosing a non-zero area on the PV diagram. This is possible only if on the PV diagram the system goes through more than one isothermal on the way back to its initial state, and thus only if it exchanges heat with heat reservoirs at more than one absolute temperature. Note that a loop cannot be closed by using only adiabatics (i.e. transformations with no heat exchange), since they do not intersect each other. Obviously, one might have a more complex machine working with more than a single cylinder. In this case, the machine cannot be trivially represented on a single PV diagram, and the argument above cannot be easily applied (it could be interesting, though, to investigate whether and how it could be extended to such engines). Thus, in classical thermodynamics, the second law appears to be a strict necessity, even a trivial one.

---

[2] Actually, this is true for all working substances, but only gases can undergo significant variations in both P and V.

1.1   Status of the second law

Even though the second law "holds the supreme position among the laws of Nature" (in Eddington's own words [10]), its position appears to be also quite peculiar. After the development of statistical physics by Maxwell, Boltzmann and other scholars, it became clear very quickly that the second law of thermodynamics could not hold unconditionally, but only statistically. Brownian motion is a well-known macroscopic example of that state of affairs, as already noted by Poincaré (see Chapter 2). In other words, the entropy of isolated systems is not forbidden to decrease, but in all processes the *probability of a continuous and macroscopically significant (able to provide usable work) entropy decrease is extremely small.*

Consider, for instance, a container separated into two compartments, A and B, by a diaphragm. Both chambers contain the same amount of ideal gas, e.g. 10^{23} particles each[3], and are at the same temperature T (Fig. 1.2). If the partition that separates chamber A from chamber B is removed, then nothing prevents the particles in chamber A from freely moving to chamber B, and vice-versa. By assuming the interaction between the particles as negligible (we are dealing with an ideal gas), the behavior of each particle is uncorrelated with that of any other particle and there is a non-zero chance that, at a certain instant of time, all the particles of both chambers occupy the sole chamber A. If one observes the system, the probability of finding a specific particle in chamber A is obviously equal to $\frac{1}{2}$; thus the probability that, at a certain instant of time, *all* the particles are in chamber A is given by,

$$P_{A\&B\to A} = \left(\frac{1}{2}\right)^{2\times 10^{23}} \simeq 10^{-6\times 10^{22}}. \qquad (1.2)$$

This probability is an incredibly tiny one, but it *is not* zero. If

---

[3]It is well known that 22.414 L of gas at standard conditions ($T = 273.15\,\mathrm{K}$, $P = 1.01355 \times 10^5\,\mathrm{Pa}$) contain 6.023×10^{23} molecules (Avogadro's Number). Hence, any macroscopic volume of gas we deal with in real life contains no less than $10^{20} - 10^{23}$ molecules.

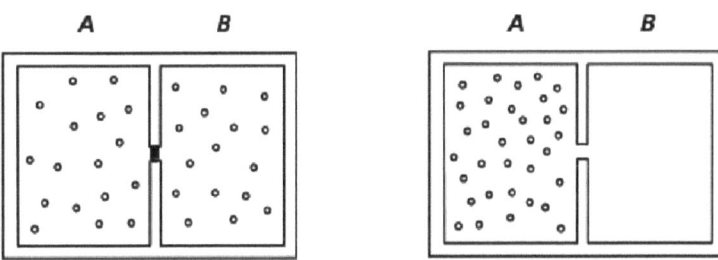

Figure 1.2: Gas-in-two-chambers thought experiment described in the text (Reprinted from [135], with permission from Elsevier)

the total length l of the two-chamber container is 1 m and the mean velocity $\langle v \rangle$ of the particles is $\sqrt{\frac{8kT}{\pi m}}$, where k is the Boltzmann's constant, T the absolute temperature of the gas and m is the mass of one particle ($\langle v \rangle$ can be derived from the Maxwell-Boltzmann velocity distribution), then the order of magnitude of the average time τ one particle takes to go from chamber A to chamber B, or vice-versa, is given by,

$$\tau = \frac{l}{\langle v \rangle} = \sqrt{\frac{\pi l^2 m}{8kT}} \simeq 5.6 \times 10^{-4} \text{ seconds,} \qquad (1.3)$$

where for m we have chosen the mass of the lightest gas molecule (hydrogen molecule) and for T the room temperature 298 K. We can consider τ as the *clock-time* at which the system of particles changes its configuration. Hence, the mean time $\langle T \rangle$ one has to wait to observe the exceptional occurrence described above (all particles in chamber A) is nearly,

$$\langle T \rangle = \frac{\tau}{P_{A\&B \to A}} \simeq 10^{6 \times 10^{22}} \text{ seconds.} \qquad (1.4)$$

Note that this time is nearly $10^{6 \times 10^{22}}$ times the estimated age of the Universe ($\sim 10^{17}$ s) since $\frac{10^{6 \times 10^{22}} \text{ s}}{10^{17} \text{ s}} \approx 10^{6 \times 10^{22}}$. Thus, no one will ever have the chance to observe that occurrence, but *this does not mean that it is forbidden by the fundamental laws of physics.*

For the sake of completeness, let us now show why the scenario above is a violation of the second law of thermodynamics: let us calculate the entropy variation $\Delta S_{A\&B \to A}$ of the two-chamber system when all particles are in chamber A and check whether it is actually negative.

As is usually done in classical thermodynamics, we calculate integral (1.1) along an isothermal compression[4] from state [**gas in volume** $A\&B$] to state [**gas in volume** A] of an ideal gas with equation of state $PV = kNT$ (k is the Boltzmann's constant and N the total number of molecules). The compression is isothermal and the internal energy U does not change ($dU = 0$). From the first law of thermodynamics $\delta Q = dU + \delta W$, we have $\delta Q = \delta W = pdV = \frac{kN}{V}dV$ and thus,

$$\Delta S_{A\&B \to A} = \int_{V_{A\&B}}^{V_A} \frac{\delta Q}{T} = \int_{V_{A\&B}}^{V_A} \frac{kN}{V}dV =$$
$$= kN \ln\left(\frac{V_A}{V_{A\&B}}\right) = -kN \ln 2 < 0, \quad (1.5)$$

being $V_{A\&B} = 2V_A$.

Consider the above experiment with only 18 molecules in each chamber. In that case, Eq. (1.4) gives $\langle T \rangle \approx 1\,\text{year}$. This means that every year, on average, this reduced system violates the second law by an amount of $|\Delta S| = k \cdot 36 \cdot \ln 2 = 3.44 \times 10^{-22}\,\frac{\text{J}}{\text{K}}$, at the most. Unfortunately, such a violation could hardly be exploited to produce usable work. Note that this is not because of its minuteness, but because we do not know exactly *when* this violation occurs. However, this difficulty has little to do with impossibility due to fundamental laws of physics. Impossibility due to fundamental laws of physics would have forbidden *any entropy decrease just from the outset.*

---

[4] As already noted, the entropy function S is a state function; and in order to calculate its variation between two equilibrium states, we can choose any reversible transformation that connects them. The choice of an isothermal compression is usually preferred, since it makes the calculation easier.

Every other known fundamental law of physics, like those of Newton's mechanics, Einstein's relativity, Maxwell's theory of electromagnetism and even the fundamental laws of quantum mechanics[5] provide an absolute and unconditional prescription on how processes should behave in Nature. For instance, Newton's laws tell us that a body that no forces act upon undergoes no acceleration; it does not tell us that the body has "a very large chance" of undergoing no acceleration. Coulomb's law tells us that two positive charges far removed from any other charge distributions will repel each other and how; not that they will repel each other "with high probability" [102]. The second law, instead, forbids some processes not absolutely, but only with *very high probability*.

1.2 Maxwell's Demon: a digression

Our thought experiment is only the most recent in a long series of two-chamber thought experiments in the context of statistical thermodynamics. The most celebrated one, in fact very similar in spirit to our own, dates back to 1867, when J. C. Maxwell introduced it to show, probably for the first time, that the second law of thermodynamics has only a *statistical validity*. Actually, he made the point more cogent by introducing what it is now widely known as "Maxwell's Demon" (Fig. 1.3). In his own words:

> [...] if we conceive of a being whose faculties are so sharpened that he can follow every molecule in its course, such a being, whose attributes are as essentially finite as our own, would be able to do what is impossible to us. For we have seen that molecules in a vessel full of air at uniform temperature are moving with velocities by no means uniform, though the mean velocity of any great number of them, arbitrarily selected, is almost exactly uniform. Now let us suppose that such a vessel is divided into two portions, A and B, by a division in which there is a small hole, and that a being, who

---

[5]Although quantum mechanics strongly hints at an intrinsic probabilistic nature of reality, its fundamental laws are in fact deterministic.

can see the individual molecules, opens and closes this hole,
so as to allow only the swifter molecules to pass from A to B,
and only the slower molecules to pass from B to A. He will
thus, without expenditure of work, raise the temperature of
B and lower that of A, in contradiction to the second law of
thermodynamics [...]. [96]

The number of molecules in A and B are the same as at first,
but the energy in A is increased and that in B diminished, [...]
and yet no work has been done, only the intelligence of a very
observant and neat-fingered being has been employed. Or in
short if the heat is the motion of finite portions of matter
and if we can apply tools to such portions of matter so as to
deal with them separately, then we can take advantage of the
different motion of different proportions to restore a uniform
hot system to unequal temperatures or to motions of large
masses. Only we can't, not being clever enough [...]. [49]

[...] *I do not see why even intelligence might not be dispensed
with and the thing be made self-acting* [emphasis added].
Moral: The 2nd law of Thermodynamics has the same degree of truth as the statement that if you throw a tumblerful
of water into the sea you cannot get the same tumblerful of
water out again. [1]

What is interesting with respect to our previous thought experiment (Fig. 1.2) is that Maxwell's thought experiment accomplishes
a violation of the second law that is, in principle, also an exploitable
violation, namely one that can produce usable work. We have just
realized, with our 36-molecule gas-in-two-chambers system, that
the violation of the second law and the exploitable violation of the
second law are not the same thing.

Maxwell's "neat-fingered being" has given rise to an incredibly
rich literature over the subsequent decades, which is still ongo-
ing and stronger than ever. Born as a simple and very effective
Gedankenexperiment to elucidate the limits of the second law, almost all the subsequent scholars forgot Maxwell's pristine intention
and focused their attention entirely on the evil being, trying to ex-

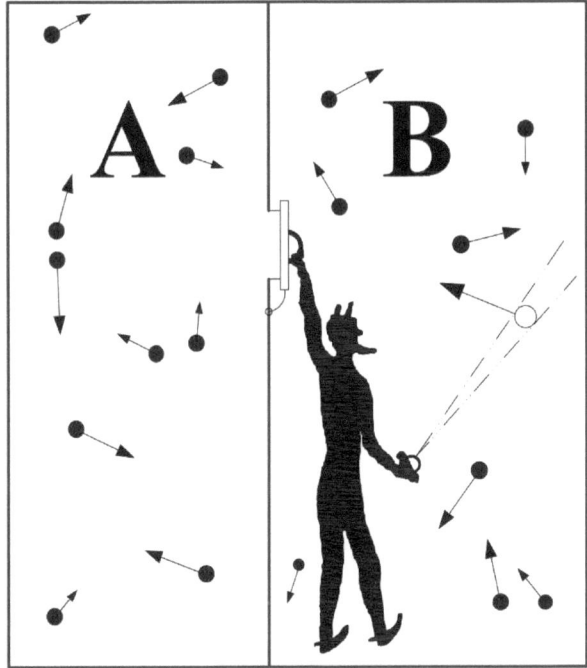

Figure 1.3: Cartoon depiction of Maxwell's demon (Adapted from [33])

orcise it; namely trying to prove the impossibility of the demon to operate in order to save the second law and to preclude the possibility of macroscopic exploitation of such a violation (production of usable work).

Smoluchowski, with his one-way valve (trapdoor) [7], and Feynman, with the ratchet and pawl analog [23], introduced a *non-sentient* version of Maxwell's demon, using pure physico-mechanical devices without the need of an "intelligent being" able to "perceive" velocities, "see" paths and "handle" molecules. In fact, the possibility of a *non-sentient* demon was anticipated by Maxwell himself (see the quotation above).

Smoluchowski and Feynman have shown that the thermal fluctuations suffered by these mechanical devices prevent any anti-entropic action, such the sorting of molecules from one vessel to the other. As a matter of fact, every mechanical device supposed to sort molecules must work at the same absolute temperature of the gas; otherwise, its action may be ascribed to the possible extraction of work from heat reservoirs at different absolute temperatures, like a standard Carnot engine, which does not count as a "regular" second law violation. Hence, the mechanical device itself must follow the same canonical distribution function associated with the temperature of its immediate surroundings. For instance, in the Smoluchowski one-way valve, molecules have an average kinetic energy of $\sim kT$ in a given direction, so the valve-trapdoor must be sensitive to energies that high, and preferably lower energies as well. However, the trapdoor has the same temperature as the molecules; it is, after all, in contact with them. This means it has fluctuations of kinetic energy of the same size as the molecules; that is, of the size of $\sim kT$. Basically, the trapdoor must be sensitive to energies of order kT, and it itself is plagued by fluctuations of order kT. It is therefore sensitive to random fluctuations, and there will be no correlation between the openings of the trapdoor and the arrival of molecules.

Other studies have attempted to investigate the original, *sentient* version of Maxwell's demon (that of intelligently operated devices). Szilard and Brillouin[6] argued that in order to achieve the entropy reduction, the intelligent being must acquire knowledge on molecule's dynamical state (position, velocity) and so must perform a measurement. Thus, they argued that the second law would be saved if the acquisition of knowledge by the demon came with a compensating entropy cost [12, 17].

In more recent years, some researchers (Bennett, Landauer and followers) have claimed that measurements can be performed without entropy costs at all. Instead, they focused their attention exclusively on the process of information erasure, needed by the sentient

---

[6]In particular, Brillouin [17] mathematically addressed in an explicit way the original form of Maxwell's demon, namely that of a "neat-fingered being" actually able to see individual molecules. He showed that in order to see the single molecule, the demon should use a (black-body) radiation more energetic (higher temperature) than that of the gas and environment, thus generating a compensating entropy increase.

demon to operate cyclically. All the information on the dynamical status of the molecules gathered and stored by the demon must be first acquired and then necessarily erased in order to operate cyclically [21, 38, 43]. According to the information erasure school, any sort of sentient demon is strictly and absolutely forbidden to violate the second law by the unavoidable entropy cost of the information erasure step. This step provides the Universe with an entropy increase greater than or equal to the alleged entropy reduction operated by the demon.

Although the connection between physical entropy and information theory is now widely recognized, its arguments appear to be either circular (typically relying on some version of the second law) or appeal to the existence of new fundamental "laws", which have nothing to do with the known fundamental physical principles (classical and quantum mechanical) that govern the behavior of matter [55, 66, 99, 114]. However, these "laws" are, in the end, a mere recasting of the second law in the lofty formalism of information theory: not an explanation of the second law by the known fundamental laws of physics, nor a proof of second law necessity [92]. A robust argument against the necessity of information acquisition (measurement) and memory erasure entropy costs to defeat Maxwell's demon is as follows.

Historically, Szilard [12], Bennett [38, 43], and several other scholars working on the same subject, have all used the Szilard one-molecule heat engine to illustrate their respective points. Szilard's heat engine works as follows. Initially, the entire volume V of a cylinder is available to a single molecule. The first step consists of placing a partition into the cylinder, dividing it into two equal chambers. In step 2, a Maxwell's demon determines which side of a partition the molecule is on, and records this result. In step 3, the partition is replaced by a piston, and the recorded result is used to couple the piston to a load upon which work W is done. The gas pressure moves the piston to one end of the container, returning the gas volume to its initial value V. In the process, the one-molecule gas has energy $Q = W$ delivered to it via heat from a constant-temperature heat reservoir. After step 3, the gas has the same volume and temperature as it had initially. The heat bath, which has transferred energy to the gas, has lower entropy than it had initially. Without some other mechanism, the second law has been violated during the cyclic process.

As a matter of fact, it is not difficult to devise a modified Szilard's heat engine that cyclically works without the need of information acquisition and/or memory erasure. Such an engine is shown in Fig. 1.4 (appended at the end of the chapter). It is made up of a movable cylinder and two pistons (the left one movable and the right one fixed). There is also a partition that can be lowered exactly in the middle of the cylinder and that can slide horizontally on a lowering rod without friction as the cylinder moves. The insertion of the partition involves no work or heat. All the mechanical parts are thought without friction, as has been shown extensively in the literature on the subject (more on this later). Initially, the entire volume V of a cylinder is available to a single molecule (step A). The behavior of the molecule is described by the equation of state $PV = kT$. Then, the partition is lowered into the cylinder, dividing it into two equal chambers (step A_1). The molecule is trapped in one of these two chambers.

Then, the movable piston is pushed infinitesimally slowly (reversibly) to position B and the one-molecule gas undergoes an isothermal compression from $V/2$ to $V/4$. The work $W_{A_1 \to B}$ externally done to the gas is equal to $kT \ln 2$, which is also equal to the heat transferred from the gas to the heat reservoir at temperature T. Note that this part of the cycle is independent of which side of the partition the molecule is on at step A_1; hence we do not need any information acquisition (with subsequent memory erasure). The final position of the movable piston is always at point B, no matter if the molecule is in the right or left chamber after step A_1. Besides, the movement of the partition can be mechanically coupled to the cyclic movement of the movable piston, thus without any need of information acquisition or memory erasure to operate the partition itself.

One may complain that the compression procedure actually depends on whether the molecule is trapped on the left or right.

Namely, if the molecule is on the left, the piston moves the whole cylinder first, with its action on the cylinder mediated by the gas pressure. If the molecule is on the right, the piston moves in unimpeded to contact the partition and then compresses the gas (moving the whole cylinder). Since the two processes appear to be slightly different, one may wonder that in order to operate the device, one actually has to know which conditions are at hand. This would mean measurement and memory erasure. Under a more careful analysis, one can easily see that the two processes are not significantly different. In both cases, there is a first phase where the device moves unimpeded until the right piston, if the molecule is on the left, or the left piston, if the molecule is on the right, touches the partition (from step A_1 to the midpoint between A_1 and B); and then a second phase in which there is true gas compression (from the midpoint between A_1 and B to step B). These two phases are physically perceived always in the same way by who/what operates the device, provided that the process is "infinitesimally slow": the first half of the compression is always equally "loose", no matter where the molecule is at the beginning of the process, while the second half is the true gas compression. If the molecule is initially on the left, the unimpeded movement of the whole cylinder before the contact of the partition with the fixed right piston is mediated by the hits of the molecule between the moving left piston and the partition.

At step B_1, the partition is raised and the cycle is completed with an isothermal expansion from $V/4$ to V (with movable piston again in position A). Now, the work $W_{B_1 \to A}$ done by the gas on the environment is equal to $kT \ln 4$, which is also equal to the heat transferred to the gas by the heat reservoir at temperature T. The net work output W_n over any cycle is then equal to $W_{B_1 \to A} - W_{A_1 \to B} = kT(\ln 4 - \ln 2) = kT \ln 2$. Moreover, the entropy variation ΔS of the entire system (engine + reservoir) is equal to $-k \ln 2$.

If we want to save the second law, some mechanisms other than information acquisition/erasure must come into play to prevent the modified Szilard engine from operating. For instance, thermal fluctuations surely plague the mechanical parts of the engine (pistons, partition and so on) [66]. The pistons must be sensitive to energy of the order of kT, the mean energy of the molecule, and they themselves are plagued by fluctuations of order kT, like the Smoluchowski one-way valve. In fact, if there were no friction, the device could operate even with arbitrarily massive pistons, partition and cylinder (massive means not instantaneously sensitive to energy of the order of kT). Without friction, even the tiny kick of a single molecule could move a massive piston/cylinder (conservation of linear momentum). However, friction cannot be eliminated, even ideally, since thermal fluctuations of the matter along the contact surface between the pistons' edge and the cylinder's walls originate an unavoidable friction force that is surely greater than the force imparted to the pistons by the molecule.

However, if such effects plague our modified Szilard engine, then the same effects must plague the original Szilard engine, both engines being mechanically similar. Hence, the appeal to information acquisition and/or memory erasure entropy costs is superfluous in defeating the Maxwell's demon, also in the original Szilard engine.

On the other hand, if information acquisition and/or memory erasure entropy costs are to be considered strictly necessary to defeat original Szilard's engine, then this means that *no other mechanisms*, like thermal fluctuations and friction, are enough to prevent its operation and the operation of every engines similar to it. However, if no mechanisms like thermal fluctuations and friction are enough to prevent the original Szilard engine from operating, then this must also apply to our modified Szilard engine, which means that our modified Szilard engine would surely violate the second law, since measurement and memory erasure, with their associated entropy costs, do not apply to it, as we have just shown. Thus, measurement and memory erasure entropy costs are again ineffective and unnecessary to defeat Maxwell's demon (this time in the instantiation of our modified Szilard engine).

As a conclusion, the appeal to measurement and memory erasure entropy costs made by Szilard and Bennett in the context of the original Szilard engine appears to be an arbitrary choice rather than a necessity in defeating the demon.

Probably the true reason why non-sentient demons cannot operate, macroscopically violating the second law and producing usable work, is the ubiquity of friction and thermal fluctuations in physical matter, the matter that inevitably constitutes both the gas and every passive device conceived to sort molecules in the gas-in-two-chambers scheme. Fluctuations make the non-sentient demon ineffective. In the end, Maxwell's thought experiment of the two-chamber vessel with a non-sentient demon becomes equivalent to our thought experiment of the two chambers connected by a wide open hole (Fig. 1.2): macroscopic violation of the second law can *only* be possible by macroscopic statistical fluctuation of molecule concentration between the chambers; the presence or the absence of a sorting non-sentient demon before the connecting hole does not make any difference at all. In addition, every sentient demon (for instance, that of Brillouin), which in order to operate needs to acquire information on the molecule (or even to erase memory), must necessarily release (exchange) energy to the gas and the environment: this is equivalent to a demon that performs work to the system and it is not a canonical Maxwell's demon, which operates "without expenditure of work", in Maxwell's own words. Thus, we have the strong feeling that every sentient demon is doubly ineffective in violating the second law in the gas-in-two-chambers scheme: first, because every mechanical part of it, which has to be "pico-metric" in order to deal with single molecules, is unavoidably plagued by thermal fluctuations[7] and friction; secondly, because every energy exchange with the gas required by the measurement process (or even by the memory erasure) may imply a further entropy increase.

---

[7]In Feynman's own words: "If we assume that the specific heat of the demon is not infinite, it must heat up. It has but a finite number of internal gears and wheels, so it cannot get rid of the extra heat that it gets from observing the molecules. Soon it is shaking from Brownian motion so much that it cannot tell whether it is coming or going, much less whether the molecules are coming or going, so it does not work.". In a recent paper, Norton [129] has provided a general result that seems to prove the unavoidability of that limitation rigorously.

1.3   Back to the second law

A critical evaluation of the literature on Maxwell's demon and, to some extent, of Maxwell's demon itself is not the main goal of this book; the interested reader is referred to [55, 66, 96, 102, 125] and references therein. Rather, our interest here is mainly in the epistemological significance of the gas-in-two-chambers scheme (with or without a sorting demon) for the status of the second law. According to the previous discussion, the only logically tenable, legitimate and more basic inference that can be drawn from the gas-in-two-chambers thought experiment (that of Maxwell but, above all, that depicted in Fig. 1.2 and described in the previous sections), is **not** that the probability of a macroscopic and exploitable violation of the second law is *always* extremely small (practically zero), and thus that the second law is safe, but that:

i) the second law is **not a necessary law**. There are no known fundamental laws of physics that absolutely forbid its violation; thus, it can be in principle macroscopically violated. So far, none of Maxwell's demon exorcisms provides first principles and fundamental laws of physics able to forbid the violation of the second law absolutely. In the end, there is no exorcism that is convincingly and beyond reasonable doubt not attributable to thermal fluctuations and friction (which are contingent causes);

ii) the probability of a macroscopic and exploitable violation of the second law is extremely small **if** one uses the gas-in-two-chambers scheme or analogs, with or without a sorting demon. As a matter of fact, thermal fluctuations make every gas-in-two-chambers scheme with a sorting (sentient or non-sentient) demon equivalent to a gas-in-two-chambers scheme with a wide open hole between the two chambers. Thus, within this scheme, macroscopic violations of the second law are possible only by macroscopic statistical fluctuation of molecule concentration between the chambers. We know that this is statistically highly improbable.

In essence, with Maxwell's and our own thought experiments, one cannot definitively prove that the second law cannot be macroscopically violated by schemes *different* form the gas-in-two-chambers ones, those for instance not involving gas, liquid or solid atoms and molecules in thermal equilibrium (whose behavior is described by the canonical distribution).

For what concerns the gas-in-two-chambers scheme described above (Fig. 1.2), the following summary inference chart holds:

1) Gas spontaneous macroscopic compression in the gas-in-two-chambers scheme	⇒	Macroscopic violation of the second law

But

2) Practical impossibility of gas spontaneous macroscopic compression in the gas-in-two-chambers scheme	⇏	Absolute macroscopic non-violability of the second law
3) Macroscopic violation of the second law	⇏	Real possibility of gas spontaneous macroscopic compression in the gas-in-two-chambers scheme, *which is actually anything but probable*

Namely, the inability of the gas-in-two-chambers scheme (with or without a sorting demon) to violate the second law macroscopically is *logically* uncorrelated with the actual possibility of second law macroscopic violation.

Inference 3) has been explicitly added since sometime people are overwhelmed by the logical fallacy that if the second law could be somehow macroscopically violated, this would automatically imply that gas spontaneous compressions in the gas-in-two-chambers scheme would actually be possible. However, since such compressions are statistically highly improbable, they argue with a sort of "inverted logic"; then, the second law cannot be macroscopically violated. These two facts, as shown at point 3), are uncorrelated in that inference direction.

What we are suggesting here is that Boltzmann's principle of statistical entropy increase[8], well represented by the high improba-

---

[8]A system approaches equilibrium because it evolves from states of lower

bility of spontaneous gas compression (see Eq. 1.2), and the possibility of macroscopic violation of the second law of thermodynamics can be both true; or, better, they are not mutually exclusive (see also [84]). Versteegh and Dieks, in a very interesting paper on the Gibbs paradox and the distinguishability of identical particles [133], pointed out that the entropy concept in thermodynamics is not completely identical to that in statistical mechanics.

As is clear from the considerations above, we have tried to argue for a possible non-necessary (contingent) nature of the second law, which leaves the door open for its violability. Obviously, contingency is a necessary but not sufficient condition for violability. Given the actual status of the second law, research aiming at the study of its violability appears to be worthwhile.

1.4   The quest for violation

Over the past 30 years, an unparalleled number of challenges has been proposed against the status of the second law. During this period, more than 50 papers have appeared in the refereed scientific literature [111]. Moreover, during the same period of time, four international conferences on the limit of the second law were also held [87, 120, 131, 142].

Obviously, not all scholars are willing to give reputable status to this line of research. For instance, Gyftopoulos and Beretta [112] wrote:

> If no challenges have been proven valid [so far], what is the motivation for pursuing exploratory research to prove that the second law is invalid? To put our question differently, why people interested in exploratory research do not try to prove that the solar system is neither geocentric nor heliocentric? Similarly why researchers do not try to prove that, in the realm of its validity, Newton's equation of motion is not correct?

---

toward states of higher probability, and the equilibrium state is the state of highest probability.

The simple answer to this question is that, as argued before, both Newton's laws of mechanics and the heliocentric theory hold a different (epistemological) status with respect to the second law of thermodynamics.

The general class of recent challenges spans classical/standard [36, 37, 39, 45, 47, 51, 75, 100, 101, 103, 104, 105, 122, 123, 126, 127], plasma [50, 52, 57], chemical [58, 69, 78, 115, 139], gravitational [70, 71, 88, 89], solid-state [90, 111, 116, 119, 132] and quantum physics [53, 54, 56, 59, 60, 63, 64, 65, 67, 68, 72, 73, 74, 77, 80, 81, 82, 83, 85, 91, 93, 94, 97, 98, 106, 110, 113, 121]. Some of these approaches appear immune to standard second law defenses; several of them account laboratory corroboration of their underlying physical processes. Others, mainly some classical/standard, gravitational and solid-state challenges, have been criticized or proved faulty beyond reasonable doubt (see, for example [40, 41, 108, 134]).

A thorough description of all these challenges is a quite hard task to accomplish and is beyond the scope of this book. The interested reader may find a very detailed review in [111], [120], [124] and [131].

In the next five chapters we shall present an exhaustive description of one of these challenges: that related to the so-called *thermo-charged capacitor*.

1.5 Concluding remarks

We have argued that we still do not have any fully cogent argument (known fundamental physical laws) that excludes the possibility of a macroscopic violation of the second law of thermodynamics in its classical formulation (Kelvin-Planck and Clausius postulates). Even Landauer's information-theoretic principle seems to fall short of the initial expectations of being the "fundamental" physical reason for all Maxwell's demons failures.

We have also provided bibliographic references for several experimental challenges that have been proposed in recent years and we mentioned the physics behind them. Nevertheless, without unambiguous experimental results (which are currently lacking) it is difficult to say whether these experiments actually violate the thermodynamic second law, even when the theory behind them appears to be sound and as yet unchallenged.

The impact of any proven success would be understandably enormous from the point of view of basic principles. However, we do not believe that, at the present stage, any successes would also have profound practical consequences. According to the theoretical modeling at our disposal, the expected power output of all these devices appears to be so minuscule that it is unthinkable to be soon able to extract usable work from environmental heat.

Surely, successful challenges would shed new light on the possible distinction between thermodynamic entropy (classical thermodynamics) and statistical entropy. As hinted at in Section 1.3, statistical entropy and the laws that govern its behavior would remain unaffected by any possible proven success of these challenges. Even if it turns out that thermodynamic second law is violable, the breakage of a glass, the mixing of milk and coffee or the mixing of two distinct gases, for instance, will always be "irreversible" processes when taken as such (namely, when not reversed in some way with the external help of any of the devices that allegedly violate the thermodynamic second law). The direction of such processes always is in the sense of increasing Boltzmann's entropy. In the case of future success of some challenges, we are sure that this last embryonic thought about such a distinction will deserve further investigation.

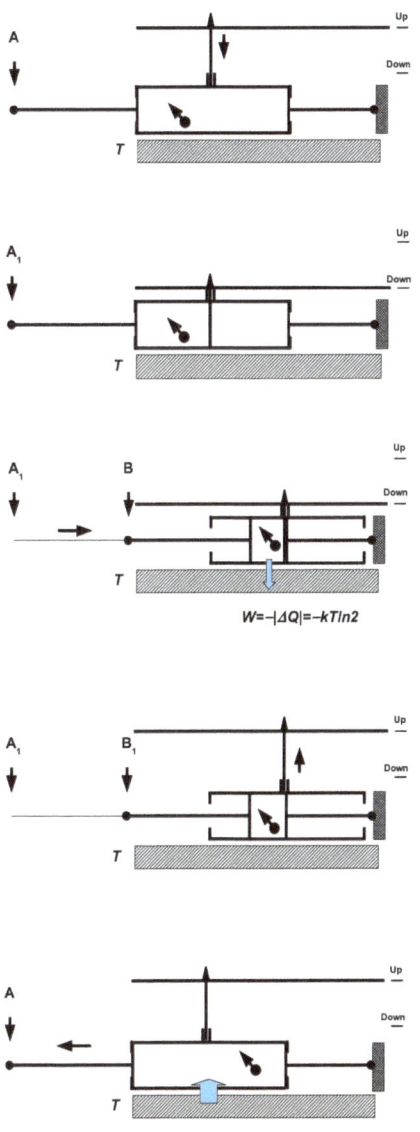

Figure 1.4: Modified Szilard heat engine described in the text (Reprinted from [135], with permission from Elsevier)

Chapter 2

Thermo-charged capacitor – Part I

In the macroscopic physical world, the second law seems authoritatively to make the difference between what is allowed and what is forbidden; to date, no experimental violation has been claimed. In the microscopic realm, however, the second law seems to be continuously violated: consider, for instance, Brownian motion or every fluctuation phenomena. About Brownian motion, Poincaré wrote [4]:

> [...] we see under our eyes now motion transformed into heat by friction, now heat changed inversely into motion, and that without loss since the movement lasts forever. This is contrary to the principle of Carnot.

With "principle of Carnot" Poincaré meant the second law of thermodynamics.

Almost every past attempt to shed light on and exploit such microscopic violations has gone hand in hand with the attempt to *rectify fluctuations*. Even the famous thought experiment of *Maxwell's demon* is in fact an idealized version of a thermal fluctuation rectifier. The main difficulties that seem to plague all these past attempts (sentient and non-sentient approaches, as described in Chapter 1) are that every macroscopic/microscopic rectifier device seems either to suffer fluctuations itself, which neutralize every usable net effect, or its functioning seems to increase the total entropy of the system (at least of the same amount of the alleged reduction) mainly because of energy dissipation and/or entropy cost in the acquisition of information needed to run the device (for sentient devices). In the previous chapter, we gave a brief historical account (along with reference to more specific publications) of second law classical challenges (starting from Maxwell, and passing through Smoluchowski, Szilard, Brillouin, until Landauer) and their alleged resolutions.

In this chapter, we give an introductory description of a different approach to microscopic rectification that poses interesting theoretical challenges: we are referring to the thermionic effect of materials emitting electrons at room temperature. More rigorously, we refer to the photoelectric emission induced by the absorption of the high-frequency tail of black-body radiation at room temperature (heat). If we succeeded in collecting all the electrons emitted by these materials as a consequence of the absorption of black-body radiation from a uniformly heated environment, then we would be able to create a macroscopic voltage out of a *single* heat reservoir.

Although this voltage difference appears to be *prima facie* an innocuous consequence of some well established physical laws, the whole process of "thermo-charging" presents quite interesting and puzzling features, which, analyzed with the tools of classical thermodynamics, appear to violate the second law and seem to provide support to the recent results presented in the literature cited in the previous chapter.

This chapter is organized as follows: in the next section, we describe what we have christened the *thermo-charged capacitor* and present an introductory mathematical analysis of its functioning: it is ultimately a vacuum capacitor with plates made of materials with different easiness in emitting electrons upon radiation absorption (work function), designed just to do what we have sketched above. In the last section, we discuss the paradoxical features of the thermo-charging process and show quantitatively why it seems

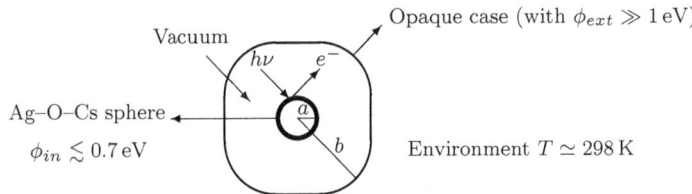

Figure 2.1: Schematic of a thermo-charged spherical capacitor with inner electrode made of Ag–O–Cs (with $\phi_{in} \lesssim 0.7\,\text{eV}$) (Reprinted from [126], with permission from Elsevier)

to be at odds with the second law of thermodynamics.

2.1 Thermo-charged spherical capacitor

In Fig. 2.1, a sketched section of a vacuum spherical capacitor is shown. The inner sphere has radius a, and it is made of a thermionic (conductive) material with relatively low work function ($\phi_{in} \lesssim 1\,\text{eV}$). In solid-state physics, the work function is the minimum thermodynamic work (i.e. energy) needed to remove an electron from a solid to a point in the vacuum immediately outside the solid surface. A low work function means relatively high thermionic emission of electrons. The outer sphere, instead, has radius b, and it is made of (conductive) material with high work function ($\phi_{ext} \gg 1\,\text{eV}$) and thus has a relatively low emission of electrons.

The capacitor is put in the dark and is submersed in black-body radiation at (room) temperature T. Both materials emit electrons after the absorption of the high-energy photons from black-body radiation; let us focus only on the outward emission from inner sphere and the inward emission from the outer sphere, the outer sphere being externally insulated. Note that no straight photoelectric emission is at work; electron emission is only due to black-body radiation.

It should be clear that in such a case, the two thermionic fluxes are different: the outward flux being greater than the inward one and the latter negligible. Actually, this is true only at the beginning of the thermo-charging process but later, with the establishing of a potential difference between the spheres, the inward and the outward *effective* flux tend to balance each other exactly. As a matter of fact, to reach our goal in this chapter it is enough to concentrate on the initial phase of the process. A detailed analysis of the balancing process is provided in the next chapter. In the following treatment, we do not explicitly handle the space charge cloud that eventually settles between the spheres, especially when the potential difference between them builds up: as matter of fact, this space charge should not considerably affect the value of the potential difference attainable across the capacitor, provided that the distance between the two spheres is not too large.

Let us describe in more detail how the device works. All the electrons emitted by the inner sphere after the absorption of the high-frequency tail of black-body radiation at room temperature are collected by the outer (very low emitting) sphere, creating a macroscopic difference of potential V. Such a process lasts until V is too high to be overcome by the kinetic energy K_e of the main fraction of emitted electrons (namely, when $K_e \lesssim eV$, where e is the charge of electron). Since the work function of the outer sphere is much greater than that of the inner one, $\phi_{ext} \gg \phi_{in}$, the reverse process, namely the electronic flux from the outer sphere to the inner one, may be neglected without sensibly affecting our results.

Here we provide an estimate of the maximum achievable value of V and an estimate of the time needed to reach such a value, given the physico-geometrical characteristics of the capacitor and the quantum efficiency curve $\eta(\nu)$ of the thermionic material of which the inner sphere is made.

As mentioned, the capacitor is in thermal equilibrium with a heat bath at room temperature (environment) and it is immersed in black-body radiation. Both spheres emit and absorb an equal amount of radiation (Kirchhoff's law of thermal radiation); thus the amount of radiation absorbed by the inner sphere is the same as that emitted according to the Planck's law of black-body radiation.

Hence, given the room temperature T, Planck's law provides us with the number distribution of photons absorbed as a function of their frequency. According to the law of photoelectric emission, the kinetic energy K_e of the emitted electron is given by the following equation,

$$K_e = h\nu - \phi, \tag{2.1}$$

where $h\nu$ is the energy of the photon of frequency ν (h is the Planck constant) and ϕ is the work function of the material. Thus, only the tail of Planck's distribution of the absorbed photons, of frequency $\nu > \nu_0 = \phi/h$, contributes to thermionic emission.

The value V of the potential difference that can be barely overtaken by an electron emitted with the absorption of a photon of frequency ν_1 is then given by the following formula,

$$eV = h\nu_1 - \phi, \tag{2.2}$$

where eV is the inter-sphere potential energy; thus,

$$\nu_1 = \frac{eV + \phi}{h}. \tag{2.3}$$

The total number of photons per unit time F_p with energy greater than or equal to $h\nu_1$, absorbed by the inner sphere in thermal equilibrium, is then given by Planck's law:

$$F_p = \frac{2\pi S}{c^2} \int_{\nu_1}^{\infty} \frac{\nu^2 d\nu}{e^{\frac{h\nu}{kT}} - 1}, \tag{2.4}$$

where S is the inner sphere surface area, c is the speed of light, k is the Boltzmann constant and T the room temperature.

If $\eta(\nu)$ is the quantum efficiency (or quantum yield) curve of the thermionic material of which the inner sphere is made, then the number of electrons F_e with kinetic energy greater than or equal to $h\nu_1 - \phi$ thermionically emitted per unit time by the inner sphere towards the outer sphere is given by,

$$F_e = 4\pi a^2 \frac{2\pi}{c^2} \int_{\nu_1}^{\infty} \frac{\eta(\nu)\nu^2 d\nu}{e^{\frac{h\nu}{kT}} - 1}, \tag{2.5}$$

where $4\pi a^2$ is the surface area of the inner sphere.

As is well known, for a vacuum spherical capacitor with inner radius a and outer radius b, the voltage V between the spheres depends upon the charge Q on each sphere as follows,

$$V = \frac{Q}{4\pi\epsilon_0} \frac{b-a}{ab}, \tag{2.6}$$

where ϵ_0 is the permittivity of the free space.

Now, we derive the differential equation that governs the thermo-charging process. In the interval of time dt the charge collected by the outer sphere is given by,

$$dQ = eF_e dt = 4\pi a^2 \frac{2\pi e}{c^2} \left(\int_{\frac{eV(t)+\phi}{h}}^{\infty} \frac{\eta(\nu)\nu^2 d\nu}{e^{\frac{h\nu}{kT}} - 1} \right) dt, \tag{2.7}$$

where we make use of Eq. (2.3) for ν_1 and $V(t)$ is the voltage at time t. Thus, through the differential form of Eq. (2.6), we have,

$$dV(t) = \frac{2\pi e}{\epsilon_0 c^2} \frac{a(b-a)}{b} \left(\int_{\frac{eV(t)+\phi}{h}}^{\infty} \frac{\eta(\nu)\nu^2 d\nu}{e^{\frac{h\nu}{kT}} - 1} \right) dt, \tag{2.8}$$

or

$$\frac{dV(t)}{dt} = \frac{2\pi e}{\epsilon_0 c^2} \frac{a(b-a)}{b} \int_{\frac{eV(t)+\phi}{h}}^{\infty} \frac{\eta(\nu)\nu^2 d\nu}{e^{\frac{h\nu}{kT}} - 1}. \tag{2.9}$$

Since our aim here is to maximize the rate at which voltage V builds up, we have to choose a and b in order to maximize the geometrical factor $a(b-a)/b$. It is not difficult to see that the maximum is reached when $a = b/2$. So we have,

$$\frac{dV(t)}{dt} = \frac{\pi e b}{2\epsilon_0 c^2} \int_{\frac{eV(t)+\phi}{h}}^{\infty} \frac{\eta(\nu)\nu^2 d\nu}{e^{\frac{h\nu}{kT}} - 1}. \tag{2.10}$$

Unfortunately, even if we were able to find a simple analytical approximation of the real quantum efficiency curve $\eta(\nu)$, the previous differential equation would hardly have a general, simple analytical solution. However, a close inspection at the Planckian integral of Eq. (2.10) gives us a hint of the asymptotic behavior of $V(t)$. Even if we do not know *a priori* $\eta(\nu)$, we know it to be a bounded function of frequency, with values between 0–1; usually, the higher is ν, the closer to 1 is $\eta(\nu)$, although this is not generally true. Therefore, independent of $\eta(\nu)$, a slight increase of $V(t)$ makes the value of the Planckian integral become smaller and smaller very rapidly. Heuristically, this suggests that $V(t)$ should show a logarithmic behavior very soon ($\frac{dV}{dt}$ tends to 0)[1].

In the remainder of this section, we provide a numerical solution of the above differential equation for the practical case of inner sphere made of Ag–O–Cs [13, 20, 35] (see Fig. 2.1). To do that, we need to adopt a further approximation, however: the adoption of a constant value for η, a sort of suitable mean value $\overline{\eta}$. The differential equation (3.15) thus becomes,

$$\frac{dV(t)}{dt} - \frac{\pi e b \overline{\eta}}{2\epsilon_0 c^2} \int_{\frac{eV(t)+\phi}{h}}^{\infty} \frac{\nu^2 d\nu}{e^{\frac{h\nu}{kT}} - 1}. \tag{2.11}$$

---

[1]Consider the following toy model of Eq. (2.10):

$$\frac{da}{dt} = K \int_a^\infty \frac{x^2 dx}{e^x - 1},$$

where K is a suitable constant. This differential equation behaves exactly as Eq. (2.10), once we choose to adopt the simplifying condition $\eta(\nu) = const$. This choice does not affect the characterization of the functional upper bound on the solution of Eq. (2.10). If the lower limit a of the Planckian integral is such that $a \gg 1$ (and in Eq. 2.10, $\frac{eV+\phi}{h}$ is actually much greater that 1 even with $V = 0$), then we have that $e^x \gg 1$ and thus can proceed with the following manipulations:

$$\frac{da}{dt} = K \int_a^\infty \frac{x^2 dx}{e^x - 1} \simeq K \int_a^\infty \frac{x^2}{e^x} dx = Ke^{-a}(2 + 2a + a^2) < Ke^{-a}e^{\frac{a}{2}} = Ke^{-\frac{a}{2}},$$

since for $a \gg 1$ we have that $(2 + 2a + a^2) < e^{\frac{a}{2}}$, Thus, the solution of the differential equation $\frac{da}{dt} = Ke^{-\frac{a}{2}}$, namely $a(x) = 2\ln\left(\frac{Kx+C}{2}\right)$, gives a functional upper bound on the asymptotic behavior of Eq. (2.10).

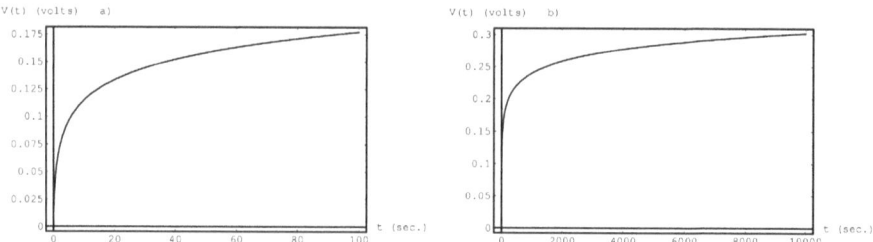

Figure 2.2: Thermo-charging process of the spherical capacitor described in the text ($\phi = 0.7\,\text{eV}$, $b = 0.20\,\text{m}$, $T = 298\,\text{K}$, and $\overline{\eta} = 10^{-5}$). These two plots show with different ranges in time-scale the behavior of $V(t)$. Plot (a) shows how only after 60 s the voltage of the capacitor becomes more than $0.15\,\text{V}$. Plot (b) tells us that the voltage of the capacitor requires some hours to approach $0.3\,\text{V}$ (Reprinted from [126], with permission from Elsevier)

A straightforward variable substitution in the integral of Eq. (2.11) allows us to write it in its final, simplified form,

$$\frac{dV(t)}{dt} = \frac{\pi e b \overline{\eta}}{2\epsilon_0 c^2} \left(\frac{kT}{h}\right)^3 \int_{\frac{eV(t)+\phi}{kT}}^{\infty} \frac{x^2 dx}{e^x - 1}. \qquad (2.12)$$

Here we adopt the following nominal values for ϕ, b, T and $\overline{\eta}$: $\phi = 0.7\,\text{eV}$, $b = 0.20\,\text{m}$, $T = 298\,\text{K}$, and $\overline{\eta} = 10^{-5}$. In order to make a conservative choice for the value of $\overline{\eta}$ we note that only black-body radiation with frequency greater than $\nu_0 = \phi/h$ can contribute to thermionic emission. This means that for the Ag–O–Cs photocathode only radiation with wavelength smaller than $\lambda_0 = hc/\phi \simeq 1700\,\text{nm}$ contributes to the emission. According to Fig. 1 in Bates [35], the quantum yield of Ag–O–Cs for wavelengths smaller than λ_0 (and thus, for frequencies greater than ν_0) is always greater than 10^{-5}.

In Fig. 2.2, the solution of this numerical test is shown. In plot (a) we can easily see how only after 60 s the voltage of the capacitor exceeds the value of $0.15\,\text{V}$. This is a macroscopic voltage indeed. Instead, plot (b) tells us that the voltage of the capacitor requires some hours to approach $0.3\,\text{V}$. Even in a more pessimistic scenario, with $\overline{\eta} = 10^{-8}$, we see that a macroscopic voltage should build up quite rapidly between the plates; see Fig. 2.3.

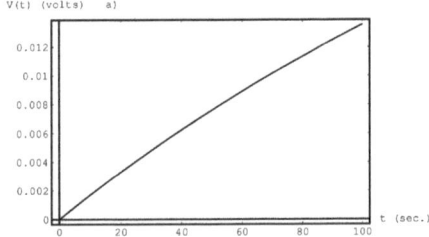
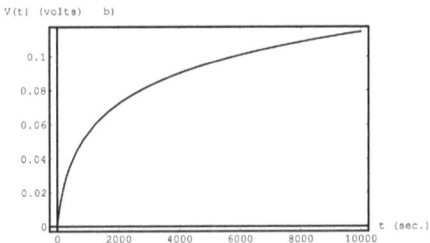

Figure 2.3: Thermo-charging process of the spherical capacitor described in the text with $\phi = 0.7\,\text{eV}$, $b = 0.20\,\text{m}$, $T = 298\,\text{K}$, and $\overline{\eta} = 10^{-8}$. These two plots show with different ranges in time-scale the behavior of $V(t)$. Plot (a) shows how only after 60 s the voltage of the capacitor is near to 0.01 V. Plot (b) tells us that the voltage of the capacitor requires some hours to become equal to 0.1 V (Reprinted from [126], with permission from Elsevier)

2.2 Discussion

At first sight, the charging process described above is a quite straightforward physical mechanism and appears almost unproblematic. One feature of the thermo-charging phenomenon, though, should catch our attention. During the charging process, the inner thermionic sphere essentially absorbs heat from the environment and releases this energy to the thermionic electrons. Such electrons fly out into to the outer sphere and impinge on it with non-zero velocity (since a non-zero fraction of them gathers their kinetic energy from the very high energetic tail of the Planck distribution of black-body radiation). When electrons come to a "stop", they release their residual kinetic energy to the outer sphere and basically heat it.

Thus, we are facing a spontaneous process involving an isolated system at uniform temperature (spherical capacitor + nearby environment), in which a part of the system (the inner sphere) absorbs heat at temperature T and transfers a fraction of this heat to the other part of the system (the outer sphere) at the same temperature. This seems to violate the second law of thermodynamics macroscopically in the Clausius formulation. In Maxwell's own words [2]:

One of the best established facts in thermodynamics is that it is impossible in a system enclosed in an envelope which permits neither change of volume nor passage of heat, and in which both the temperature and the pressure are everywhere the same, to produce any inequality of temperature or of pressure without the expenditure of work.

As a matter of fact, if Q_i is the energy absorbed from the environment by the inner sphere, U the energy stored in the electric field between the spheres ($U = \frac{1}{2}CV^2$, where $C = \frac{4\pi\epsilon_0 ab}{b-a}$ is the capacitance of the spherical capacitor), and Q_f the residual energy transferred to the outer sphere as heat via the flying electrons (according to the first law of thermodynamics $Q_f + U = Q_i$, thus $Q_i > Q_f$), then the Clausius entropy variation of the whole system, as a rough estimate, amounts to,

$$\Delta S_{tot} \simeq -\frac{Q_i}{T} + \frac{Q_f}{T} < 0. \tag{2.13}$$

In order to make the above result more striking, let us consider the following analog in classical mechanics/thermodynamics: a boulder of mass m rests at the bottom of a valley, below a hill of height h, all the system at constant temperature T. Suddenly, the boulder spontaneously absorbs an amount Q_1 of heat (energy) from the environment and spontaneously starts to climb the hill at decreasing velocity (since the initial energy is gradually transformed into gravitational potential energy). Near the top of the hill, the boulder hits a wall and finally stops. The friction experienced during the hit against the wall makes the boulder release to the environment an amount Q_2 of heat, obviously smaller than Q_1. According to the first law of thermodynamics, we have: $Q_1 - Q_2 = mgh$, where mgh is the gravitational potential energy variation of the boulder from the valley to the top of the hill.

Now, the total Clausius entropy variation is:

$$\Delta S_{tot} = -\frac{Q_1}{T} + \frac{Q_2}{T} = -\frac{mgh}{T} < 0. \tag{2.14}$$

The behavior of the boulder-environment system is practically the same as that of our electrons-environment system, and is undoubtedly puzzling from the point of view of the second law of thermodynamics.

Furthermore, the unbalanced behavior of the electrons *just after* the emission from the Ag–O–Cs coating is governed by the mechanical/ballistic laws of motion (the same governing the boulder behavior in the boulder-environment analog) and not by the canonical distribution that describes systems in thermal equilibrium, $p(\mathbf{x}, \mathbf{p}) = \frac{e^{-E(\mathbf{x},\mathbf{p})/kT}}{Z}$, where $E(\mathbf{x}, \mathbf{p})$ is the energy of the system, Z is the normalizing partition function, and the multi-component $\mathbf{x}$ and $\mathbf{p}$ are generalized configuration and momentum coordinates. Obviously, since the thermo-charged capacitor is open-circuited, a dynamical equilibrium of electrons (space charge inside the capacitor) is eventually reached, and this final state should be describable by a suitable canonical distribution. The functioning of our device is not representable on the PV diagram, which otherwise would have implied the impossibility of a second law violation, as observed in Chapter 1.

Since the thermo-charging process is spontaneous, the variation of the Gibbs free energy of the system $\Delta G = \Delta H - T\Delta S$ must be negative. The enthalpy variation for constant pressure systems is given by $\Delta H = \Delta E + \Delta W$, where ΔE is the system internal energy variation and ΔW is the work produced. ΔH is equal to 0 since, from the first law of thermodynamics, $\Delta H = \Delta Q = 0$ (we are dealing with an isolated system – capacitor plus nearby environment – and the total ΔQ is equal to zero), and thus we have $\Delta G = -T\Delta S > 0$. This is another way to see the paradoxical behavior of our system.

One possible objection to this result could be that some other physical changes take place in the system during the process that may cancel out the apparent entropy decrease. Surely, during the charging process, an electric field, and thus an electric potential, is generated inside the capacitor where none existed before. Thus, one could attribute to the creation of the electric field an entropy increase greater than the entropy decrease due to the cooling of the inner sphere and warming of the outer one.

A problem with this explanation follows directly from the logic behind the entropy variation analysis done for an ideal Carnot engine: the electric field represents a sort of work produced and stored in potential form by our capacitor, much like what happens with a standard mechanical Carnot engine whose work is stored in a lifted weight. Therefore, we are not able to see why for a classical Carnot engine (and for classical thermodynamics) the work W is not taken into account in the evaluation of the total entropy variation (unless it is transformed into heat, but this is not the case here); while for our thermo-charged capacitor, the macroscopic stored energy (work) U should be taken into account for the calculation of the total ΔS. In order to save the validity of the second law of thermodynamics, one needs to find other sources of positive entropy during the charging process.

These results do not seem to depend upon the particular value of the work function of the outer sphere, provided that $\phi_{ext} \gg \phi_{in} \simeq 1\,\text{eV}$. One may suppose that when the electrons fly out into the sphere with higher work function (i.e. the outer sphere), they may undergo a reduction of some sort of potential energy related to the work function of that sphere, thus releasing an energy greater than the mere residual kinetic energy they have, just before impinging on the sphere. This would yield $Q_f > Q_i$ in Eq. (2.13), eventually giving $\Delta S > 0$. But what about the conservation of energy?

As a matter of fact, if at some point of the charging process we decide to neutralize both spheres by putting them into contact at the same time with two distinct huge chunks of neutral materials, respectively of the same composition of the two spheres (in order to avoid problems with junction potentials) and at the same temperature T, this process will release further energy through the Joule effect of the discharge (as a matter of fact, each sphere disperses its charge into the huge chunk of neutral material). This energy is equal to U, the energy stored in the thermo-generated electric field. The energy balance ΔE_{tot} of the entire process, thermo-charging and neutralization, thus gives: the total absorbed energy amounts to Q_i, the released energy amounts to $Q_f + U$ but, according to the above alleged resolution of the entropy paradox, $Q_f > Q_i$ and thus $\Delta E_{tot} = Q_f + U - Q_i > U > 0$. Namely, we would have spontaneous energy production inside an isolated system. Therefore, the argument above cannot be an acceptable explanation of the paradox.

In 1952 Sommerfeld [18], trying to show how a *quasi-static* process does not always mean a *reversible* one, took as an example the slow discharge of a charged capacitor through a very large resistance submersed in a heat reservoir: the discharge will take place by an arbitrarily small current, and negligible disturbance of electrostatic equilibrium; thus, such a process is *quasi-static*, but not *reversible*. In fact, the reverse process, he said, is not allowed by the second law [79].

To our knowledge, the results illustrated in this chapter represent an easily understandable and not so trivially refutable challenge to the Clausius formulation of the second law (see Eq. 2.13). At this stage, however, it is not completely clear how this alleged challenge could be made exploitable; namely, how our capacitor could be transformed into a device able to produce usable net work *cyclically*[2] out of a single heat bath (i.e. challenging the Kelvin-Planck formulation of the second law). This is actually what we shall explore in the next chapter. In the next chapter, we shall deepen and extend the previous analysis in order to find an efficient route from the above "Clausius violation" to a "Kelvin-Planck violation" of the second law.

---

[2]For instance, a DC current or an electro-mechanical work like in [90, 116, 119, 124]. We cannot reasonably rely on the inter-sphere heat flux alone, since it is microscopic and can hardly be exploited thermodynamically/mechanically with an ancillary standard heat engine.

Chapter 3

Thermo-charged capacitor – Part II

In the previous chapter we introduced the concept of a vacuum capacitor spontaneously charged by harnessing the heat absorbed from a single thermal reservoir at room temperature. For brevity, we baptized it the *thermo-charged capacitor*. Furthermore, we presented a preliminary mathematical description of its basic functioning along with a Clausius entropy variation analysis: the macroscopic behavior of the thermo-charged capacitor appears to violate the Clausius formulation of the second law of thermodynamics.

In the present chapter, we investigate the practical possibility of exploiting thermo-charged capacitors as voltage/current generators. In Section 3.1, we review the setup of the thermo-charged capacitor and its mathematical model, extending those presented in the previous chapter. Here we drop an apparently crucial simplification and show that the physical process presented in Chapter 2 is robust against such a simplification.

In Section 3.2, we show that if very weak provisos on the physical characteristics of the capacitor are fulfilled, a non-zero current should flow across the device, allowing the generation of potentially usable voltage, current and electric power out of a single thermal source at room temperature. Preliminary numerical results show

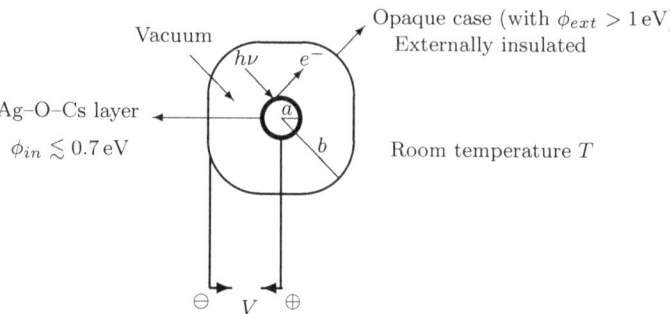

Figure 3.1: Schematic of the thermo-charged spherical capacitor with $\phi_{ext} > \phi_{in}$ and $\phi_{ext} > 1$ eV (Reprinted from [127], with permission from Elsevier)

that the power output is tiny but non-zero. If it were possible to verify such results unambiguously experimentally, we would have an experimental violation of the second law in the Kelvin-Planck formulation, which parallels the alleged violation in the Clausius formulation discussed in Chapter 2. In other words, we would have a sort of *thermo-voltaic cell*.

3.1   More on the thermo-charged spherical capacitor

For the sake of clarity and in order to make the reading easier, we decide to reproduce here (Fig. 3.1) the sketch of the vacuum spherical capacitor introduced in Chapter 2. The physical specifications are the same as those given there. There are two important differences. First, we are interested in the possibility of exploiting the potential difference between the inner and the outer sphere, thus two terminals (those indicated with symbols ⊖ and ⊕) have been added. This means that now the inner sphere is made of the same conductive material as the outer one, but is coated with a layer of semiconductor Ag–O–Cs. The ⊕ terminal is only connected to this metallic substrate. It is within the scope of this chapter to show that an exploitable potential difference actually builds up also between the external terminals $\ominus$ and $\oplus$.

Moreover, we extend the mathematical analysis of Chapter 2 to include into the equations the thermionic emission of the outer sphere (toward the inner one): here we have $\phi_{ext} > \phi_{in}$ and no longer $\phi_{ext} \gg \phi_{in}$.

An issue that needs to be addressed and that is crucial for what follows is that we now have a contact surface between the Ag–O–Cs layer and the conductive material of the inner sphere. The behavior of this contact surface must be analyzed. The contact surface between the inner metallic plate and the Ag–O–Cs layer is a well-known case of a Schottky rectifying junction (metal/n–type semiconductor). When two materials (in our case, a metal and a semiconductor) are physically joined, so as to establish a uniform chemical potential, that is a single Fermi level, some electrons are transferred, due to diffusive forces, from the material with the lower work function ϕ_1 (Ag–O–Cs) to the material with the higher work function ϕ_2 (metal). As a result, a contact potential V_c is established across the contact surface such that $eV_c = \phi_2 - \phi_1$. The junction is the region where, at equilibrium, a balance between electrostatic and diffusive (thermally driven) forces is attained. Electrostatic forces tend to reestablish electrical neutrality (by pulling the electrons back) against diffusive forces.

The energy band profiles of the semiconductor-metal junction at equilibrium are shown in Fig. 3.2. The feature that counts for the functioning of our device is that the energy levels of the vacuum for the Ag–O–Cs layer and for the metallic substrate are preserved[1]

---

[1] This is also what one should expect according to simple logic. It is difficult to believe that a micro-metric layer along the contact region can affect the intrinsic physical properties (e.g. work function) of the whole samples. If one accepts that work functions change far from the contact region as a consequence of the contact, how do they change? Is the greater work function to be lowered to the same value of the lower one? And why not the other way around? Or do they reach an intermediate value? If so, which? Moreover, if both work functions of the two samples in contact were equal to an intermediate value after the contact, then the contact potential V_c across the contact region would

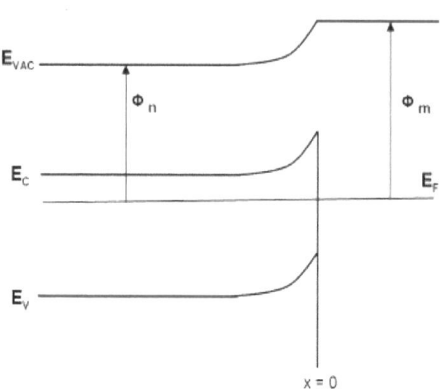

Figure 3.2: Band profiles of semiconductor (n) – metal (m) junction at equilibrium; ϕ_n and ϕ_m (with $\phi_n < \phi_m$) are the work functions of semiconductor and metal, respectively. ($x = 0$ indicates the contact between surfaces) (Reprinted from [127], with permission from Elsevier)

far from the junction [26].

This means that whenever an electron is extracted from the Ag–O–Cs layer to the vacuum (toward the outer sphere), it is always at the cost of $\approx 0.7\,\mathrm{eV}$ and, as a consequence, the bulk of the Ag–O–Cs layer starts to charge up positively. Therefore, a sort of "external" reverse bias starts to form across the junction and, although the junction is a rectifying one, a tiny current of electrons begins to flow from the metallic inner sphere to the Ag–O–Cs layer in order to reestablish the equilibrium (constant contact potential). That current flow is known as *reverse bias leakage current* (RLC). Its amplitude is influenced by many factors; among others, the thickness of the junction region (depletion region), temperature, cross-sectional area and impurities of the junction. Also the amplitude of the reverse bias influences the intensity of RLC, although when the reverse bias is below the *breakdown voltage* of the junction (usually 10s or 100s V), the current changes little with bias.

Depending on the material used, the *reverse leakage current*

---

be equal to 0 *V*. However, we experimentally know that it is not zero.

density j_0, namely RLC per unit surface, spans[2] from 10^{-6} to 10^{-9}A/cm^2 for reverse biases of the order of V or 0.1 V. As will be clear later, if we suitably increase the contact surface between the inner metallic sphere and the Ag–O–Cs layer, by simply increasing the surface S_a of the inner sphere, RLC can be made of the order of 10^{-3}A or greater, RLC being equal to $j_0 S_a$. We can also be favored by junction impurities: they usually transform rectifying junctions into ohmic ones. All these issues will be treated more thoroughly in Section 3.2.

In almost all scientific texts (both textbooks and scientific papers; see, for instance, [95]) it is stated that, when two materials with different work functions are physically joined at one end, a voltage drop ΔV builds up not only just across the contact surface, but also instantaneously between the surfaces at the free ends of the materials, where charges also accumulate, as an instant consequence of their being joined at one end (see Fig. 3.3). To the author's knowledge, no textbook or published paper on the subject clearly explains where these charges come from. This voltage drop is said to be of the same magnitude of the contact potential V_c. All this is usually explained by appealing to a supposedly straightforward application of the Kirchhoff's second (loop) rule.

If this were true, then in our device it would prevent any electron emitted by the inner layer (one free end) from reaching the outer sphere (the other free end) when the two spheres are electrically shorted through an external resistor so as to establish a circuit similar to that in Fig. 3.3. Imagine J-I to correspond to the contact junction between the Ag–O–Cs layer and the metallic substrate of the inner sphere, while J-II corresponds to the vacuum space between the Ag–O–Cs layer (B) and the outer sphere (A). The reader should imagine the bulk of A in Fig. 3.3 to correspond

---

[2]See, for example, [25, 44, 76, 86] for some specific types of Schottky and n–p junctions. Values greater than 10^{-6}A/cm^2 and lower than 10^{-9}A/cm^2 are also possible with the same reverse bias, depending on the materials, preparation, junction impurities and surface treatments. Usually, applied researchers and industry desire to lower the reverse leakage current in order to magnify the rectifying properties of the junction for electrical and electronic applications. Here, instead, we have opposing needs.

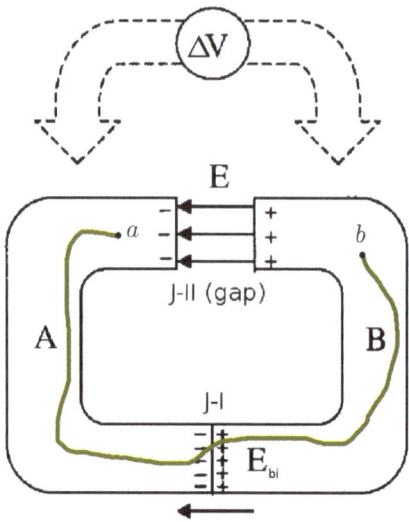

Figure 3.3: Circuit of two connected materials A and B in a vacuum. This scheme holds for metal to metal and metal to semiconductor junctions. Work functions are such that $\phi_A > \phi_B$. J-I is the physical junction while J-II is the space between the free ends of the materials (gap). The colored line is the path γ from point a to point b inside the bulk of the device. According to almost all the textbooks, a voltage drop and an electric field instantaneously build up at the free ends J-II as an instant consequence of the physical junction of A and B at one end (J-I). The analysis made in the text shows that ΔV, and thus $\mathbf{E}$, is equal to zero (Adapted from [95])

to the outer sphere plus the metallic wire connecting the outer with the inner sphere.

In that case, in order to reach the outer sphere, any electron escaping the inner layer would need the same energy as that to escape the outer sphere to reach the inner one. The inner electron needs an energy equal to $\phi_{in} + e\Delta V$, since it must be ejected (requiring energy ϕ_{in}) and then has to overcome the potential drop ΔV (energy equal to $e\Delta V$). The outer electron needs only the energy to be ejected ϕ_{ext}. However, according to the alleged origin and magnitude of ΔV, we have $e\Delta V = eV_c = \phi_{ext} - \phi_{in}$ and thus the inner electron needs $\phi_{in} + e\Delta V = \phi_{ext}$.

Such a situation would undoubtedly be one of equilibrium. However, we have explicitly shown in [128, 134] that no electric field, and thus no voltage drop, builds up between the surfaces at the free ends of two materials with different work functions when these materials are physically joined at one end. In those papers, we presented three arguments: the first two were mainly heuristic, while the third was more theoretical and consisted of an explicit application of the *path-independence* law and/or Kirchhoff's loop rule. We shall return to this in the next chapter. Here we sketch a more direct approach to reach the same conclusion.

Consider a test electronic charge e conveyed along a path γ in the device bulk of Fig. 3.3 from any point a of material A to any point b of material B and crossing the J-I physical junction. The system is intended to be at equilibrium; we now ignore any effect related to any thermionic emission whatsoever (thermionic emission switched off). The potential difference ΔV_{ba} is related to the work the charge undergoes from a to b from *all* the forces present along the path by the following well-known relation,

$$\Delta V_{ba} = -\frac{1}{e} \int_a^b \mathbf{F}_{int} \cdot d\vec{\gamma}, \qquad (3.1)$$

where $\mathbf{F}_{int}$ is the force field present along the path.

At equilibrium, the only two regions where forces are allowed to be non-null are the J-I and J-II regions. An electric field elsewhere in A and B (other than in the contact region) would generate a current, which contradicts the assumption of equilibrium. When the test charge e crosses J-I, it is subject to the built-in electric field force $e\mathbf{E}_{bi}$ ($\mathbf{E}_{bi}$ is related to the junction contact potential V_c as $V_c = x_d\mathbf{E}_{bi}$, where x_d is the width of the depletion region) and to the diffusion force $\mathbf{F}_{diff}$. This latter "force" is thermally driven and is responsible for the establishment of the contact potential at J-I. We know that at equilibrium $e\mathbf{E}_{bi} = -\mathbf{F}_{diff}$ and that $\mathbf{F}_{diff}$ is different from zero and constantly present, otherwise $\mathbf{E}_{bi}$ (and V_c) would soon drop to zero; thus,

$$\Delta V_{ba} = -\frac{1}{e} \int_a^b \mathbf{F}_{int} \cdot d\vec{\gamma} = -\frac{1}{e} \int_{\mathbf{J\text{-}I}} (e\mathbf{E}_{bi} + \mathbf{F}_{diff}) \cdot d\vec{\gamma} = 0. \quad (3.2)$$

This means that the potential difference between any point a of material A and any point b of material B is always zero. Thus, it must be zero also between the free surfaces of A and B within the J-II gap.

We now refine the estimate of the maximum achievable voltage and the estimate of the time needed to reach such a value, taking into account not only the physico-geometrical characteristics of the capacitor and the quantum efficiency curve $\eta_{in}(\nu)$ of the thermionic material Ag–O–Cs (as done in Chapter 2), but also the thermionic emission and the quantum efficiency curve $\eta_{ext}(\nu)$ of the outer sphere. Again, for the sake of clarity in the derivation of the main result and in order to make this chapter as self-contained as possible, we reproduce almost verbatim some of the preliminaries presented in Chapter 2.

As before, the capacitor is placed in a heat bath at room temperature and it is immersed in black-body radiation. Both spheres, at thermal equilibrium, emit and absorb an equal amount of radiation (Kirchhoff's law of thermal radiation); thus the amount of radiation absorbed by the spheres is the same as that emitted by them according to Planck's law of black-body radiation. Therefore, given the room temperature T, Planck's law provides us with the number distribution of photons absorbed as a function of their frequency.

According to the law of photoelectric emission, the kinetic energy K_e of an electron emitted by the material after the absorption of a photon of energy $h\nu$ $(> \phi)$ is given as follows:

$$K_e = h\nu - \phi, \quad (3.3)$$

(h is the Planck constant and ϕ is the work function of the material). Thus, only the tail of Planck's distribution of the absorbed photons, of frequency $\nu > \nu_0 = \phi/h$, contributes to thermionic emission.

An electron emitted by the inner sphere, after the absorption of a photon of frequency ν_1, is able to reach the outer sphere with zero velocity only when,

$$eV = h\nu_1 - \phi_{in}, \tag{3.4}$$

where V is the inter-sphere voltage reached so far. On the other hand, an electron escaping the outer sphere does not have to overcome an opposing voltage in order to reach the inner sphere, and thus the following holds,

$$h\nu_2 = \phi_{ext}. \tag{3.5}$$

Hence, we have these two useful relations,

$$\nu_1 = \frac{eV + \phi_{in}}{h}, \tag{3.6}$$

$$\nu_2 = \frac{\phi_{ext}}{h}, \tag{3.7}$$

which provide, for a given V, the minimum frequency of radiation with sufficient energy to move an electron from one sphere to the other.

The total number of photons per unit time F_p absorbed in thermal equilibrium by the inner sphere with energy greater than or equal to $h\nu_1$ is given by Planck's law,

$$F_p = \frac{2\pi S_a}{c^2} \int_{\nu_1}^{\infty} \frac{\nu^2 d\nu}{e^{\frac{h\nu}{kT}} - 1}, \tag{3.8}$$

where S_a is the inner sphere surface area (c is the speed of light, k is the Boltzmann constant and T the room temperature).

If $\eta_{in}(\nu)$ is the quantum efficiency (or quantum yield) curve of the Ag–O–Cs thermionic layer, the number of electrons per unit time F_{in} emitted by the inner sphere towards the outer one with kinetic energy greater than or equal to $h\nu_1 - \phi$ is given by,

$$F_{in} = 4\pi a^2 \frac{2\pi}{c^2} \int_{\nu_1}^{\infty} \frac{\eta_{in}(\nu)\nu^2 d\nu}{e^{\frac{h\nu}{kT}} - 1}, \tag{3.9}$$

where $4\pi a^2$ is the surface area of the inner sphere.

Following the same reasoning, the number of electrons per unit time F_{ext} emitted by the outer sphere and collected by the inner one is given by the following similar relation,

$$F_{ext} = \frac{2\pi S_{eff}}{c^2} \int_{\nu_2}^{\infty} \frac{\eta_{ext}(\nu)\nu^2 d\nu}{e^{\frac{h\nu}{kT}} - 1}. \tag{3.10}$$

In Eq. (3.10), it is not easy to characterize the multiplicative factor S_{eff} associated with the surface area: before the charging process starts, S_{eff} is equal to the inner sphere surface area $4\pi a^2$, but as soon as the inner sphere charges up positively, we expect that the effective surface area increases due to a sort of electrostatic focusing effect (similar to the gravitational focusing effect). Moreover, it is not easy to mathematically model such a phenomenon, since the effective area of the inner sphere depends upon the velocity of the single electron flying toward it. Here we decide to be extremely conservative and choose S_{eff} equal to the surface area of the outer sphere. As a matter of fact, as we shall see, the following results are practically independent of the choice of any reasonable value of S_{eff}.

For a vacuum spherical capacitor with internal radius a and external radius b, the voltage V between the spheres and the charge Q on each of them are related to one another by the following well-known equation,

$$V = \frac{Q}{4\pi\epsilon_0} \frac{b - a}{ab}, \tag{3.11}$$

(ϵ_0 is the permittivity of the free space).

Now, we derive the differential equation that governs the thermo-charging process in a capacitor with both spheres emitting electrons. In the interval of time dt, the charge collected by the outer sphere is given by,

$$dQ = e(F_{in} - F_{ext})dt = \frac{2\pi e}{c^2} \left(4\pi a^2 \int_{\frac{eV(t)+\phi_{in}}{h}}^{\infty} \frac{\eta_{in}(\nu)\nu^2 d\nu}{e^{\frac{h\nu}{kT}} - 1} + \right.$$

$$\left. - 4\pi b^2 \int_{\frac{\phi_{ext}}{h}}^{\infty} \frac{\eta_{ext}(\nu)\nu^2 d\nu}{e^{\frac{h\nu}{kT}} - 1} \right) dt, \quad (3.12)$$

where we make use of Eqs. (3.6) and (3.7) for ν_1 and ν_2, and $V(t)$ is the voltage drop at time t. Hence, through the differential form of Eq. (3.11), we have,

$$dV(t) = \frac{2\pi e}{\epsilon_0 c^2} \left(\frac{a(b-a)}{b} \int_{\frac{eV(t)+\phi_{in}}{h}}^{\infty} \frac{\eta_{in}(\nu)\nu^2 d\nu}{e^{\frac{h\nu}{kT}} - 1} + \right.$$

$$\left. - \frac{b(b-a)}{a} \int_{\frac{\phi_{ext}}{h}}^{\infty} \frac{\eta_{ext}(\nu)\nu^2 d\nu}{e^{\frac{h\nu}{kT}} - 1} \right) dt, \quad (3.13)$$

or

$$\frac{dV(t)}{dt} = \frac{2\pi e}{\epsilon_0 c^2} \left(\frac{a(b-a)}{b} \int_{\frac{eV(t)+\phi_{in}}{h}}^{\infty} \frac{\eta_{in}(\nu)\nu^2 d\nu}{e^{\frac{h\nu}{kT}} - 1} + \right.$$

$$\left. - \frac{b(b-a)}{a} \int_{\frac{\phi_{ext}}{h}}^{\infty} \frac{\eta_{ext}(\nu)\nu^2 d\nu}{e^{\frac{h\nu}{kT}} - 1} \right). \quad (3.14)$$

Since our aim is to maximize the rate at which voltage V builds up, we have to choose a and b such that they maximize the geometrical factor $a(b-a)/b$: the rightmost integral of Eq. (3.14) almost always has a numerical value smaller than that of the leftmost integral by several orders of magnitude, at least at the beginning of the charging process; therefore we concentrate on the maximization of the factor $a(b-a)/b$ alone. It is not difficult to see that the maximum is reached when $a = b/2$. So we have,

$$\frac{dV(t)}{dt} = \frac{\pi e b}{2\epsilon_0 c^2} \left(\int_{\frac{eV(t)+\phi_{in}}{h}}^{\infty} \frac{\eta_{in}(\nu)\nu^2 d\nu}{e^{\frac{h\nu}{kT}} - 1} - 4 \int_{\frac{\phi_{ext}}{h}}^{\infty} \frac{\eta_{ext}(\nu)\nu^2 d\nu}{e^{\frac{h\nu}{kT}} - 1} \right).$$

$$(3.15)$$

As in the previous chapter, in the remainder of this section we provide a numerical solution of the above differential equation for some realistic values of the physical parameters involved. Accordingly, we need to introduce the same approximation made in Chapter 2: it consists of the adoption of a constant value for the functions $\eta(\nu)$, a sort of suitable mean values $\overline{\eta}$.

The differential equation (3.15) thus becomes,

$$\frac{dV(t)}{dt} = \frac{\pi e b}{2\epsilon_0 c^2} \left(\overline{\eta}_{in} \int_{\frac{eV(t)+\phi_{in}}{h}}^{\infty} \frac{\nu^2 d\nu}{e^{\frac{h\nu}{kT}} - 1} - 4\overline{\eta}_{ext} \int_{\frac{\phi_{ext}}{h}}^{\infty} \frac{\nu^2 d\nu}{e^{\frac{h\nu}{kT}} - 1} \right).$$

$$(3.16)$$

Moreover, a straightforward variable substitution in the integrals of Eq. (3.16) allows us to write the equation in its final simplified form,

$$\frac{dV(t)}{dt} = \frac{\pi e b}{2\epsilon_0 c^2} \left(\frac{kT}{h} \right)^3 \left(\overline{\eta}_{in} \int_{\frac{eV(t)+\phi_{in}}{kT}}^{\infty} \frac{x^2 dx}{e^x - 1} - 4\overline{\eta}_{ext} \int_{\frac{\phi_{ext}}{kT}}^{\infty} \frac{x^2 dx}{e^x - 1} \right).$$

$$(3.17)$$

Here we provide an exemplary numerical solution of Eq. (3.17), adopting the following nominal values for ϕ_{in}, ϕ_{ext}, b, T and $\overline{\eta}_{in}$ and $\overline{\eta}_{ext}$: $\phi_{in} = 0.7\,\text{eV}$, $\phi_{ext} = 4.0\,\text{eV}$, $b = 0.2\,\text{m}$, $T = 298\,\text{K}$, $\overline{\eta}_{in} = 10^{-5}$, and $\overline{\eta}_{ext} = 1$.

Even in the present analysis, a conservative choice is made for the value of $\overline{\eta}_{in}$: we note that only black-body radiation with frequencies greater than $\nu_0 = \phi_{in}/h$ can contribute to thermionic emission. This means that for the Ag–O–Cs photocathode, only radiation with wavelength smaller than $\lambda_0 = hc/\phi_{in} \simeq 1700\,\text{nm}$ contributes to the emission. According to Fig. 1 in Bates [35], the

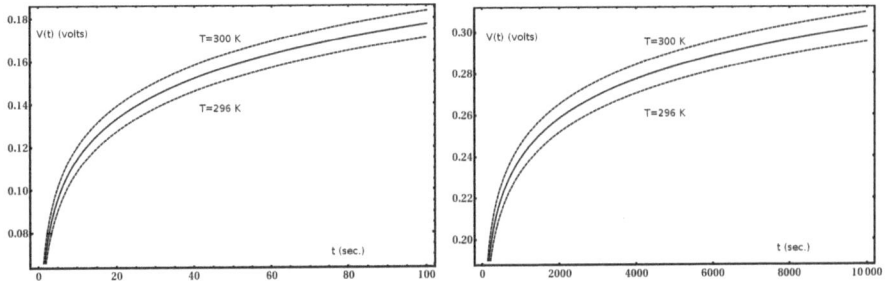

Figure 3.4: Thermo-charging profiles for the spherical capacitor described in the text ($\phi_{in} = 0.7\,$eV, $\phi_{ext} = 4.0\,$eV, $b = 0.2\,$m, $T = 298\,$K, $\overline{\eta}_{in} = 10^{-5}$ and $\overline{\eta}_{ext} = 1$). These two plots show with different ranges in time-scale the behavior of $V(t)$. Charging profiles for $T = 300\,$K and for $T = 296\,$K are also shown (Reprinted from [127], with permission from Elsevier)

quantum yield of Ag–O–Cs for wavelengths smaller than λ_0 (and thus, for frequencies greater than ν_0) is always greater than 10^{-5}.

Concerning $\overline{\eta}_{ext}$, its value is related (as for the value of ϕ_{ext}) to the discharging process due to counter emission from the outer sphere. As a matter of fact, thermionic counter-emission can be kept extremely low with the value of ϕ_{ext} high compared to ϕ_{in}, and also by choosing for the outer sphere a suitable metallic material with very low $\overline{\eta}_{ext}$; namely $\overline{\eta}_{ext} \ll \overline{\eta}_{in}$.

There are also other possible sources of counter-emission, e.g. secondary electron emission [14]; but as far as we know, they are always smaller in magnitude than the charging emission (and can be kept very low with a particular design, choice of material and surface texture of the outer sphere). In any case, counter-emission should only retard the achievement of the equilibrium voltage V, not preclude it. Any such delay in time can be easily modeled with the same Eq. (3.17), simply by suitably tuning the numerical values of ϕ_{in}, ϕ_{ext}, $\overline{\eta}_{ext}$ and $\overline{\eta}_{in}$.

In order to obtain the least involved sample solution of Eq. (3.17), we adopt the very conservative choice of $\overline{\eta}_{ext} = 1$. In Fig. 3.4 the solution of the above numerical test is shown. In Fig. 3.5 we calculated the solution in a more pessimistic scenario with $\overline{\eta}_{in} = 10^{-8}$. For the sake of completeness, in Fig. 3.4 and Fig. 3.5, the charging

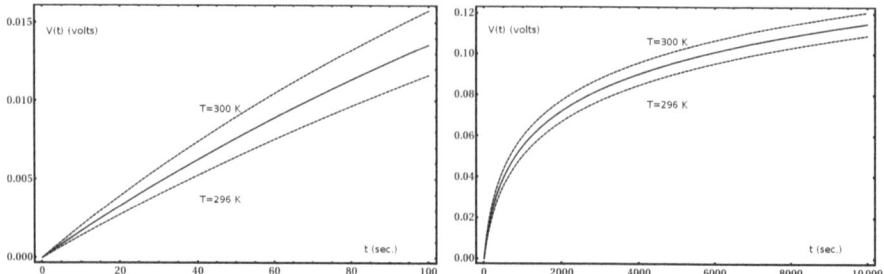

Figure 3.5: Thermo-charging profiles for the spherical capacitor described in the text with $\phi_{in} = 0.7\,\text{eV}$, $\phi_{ext} = 4.0\,\text{eV}$, $b = 0.2\,\text{m}$, $T = 298\,\text{K}$, $\overline{\eta}_{in} = 10^{-8}$ and $\overline{\eta}_{ext} = 1$. These two plots show with different ranges in time-scale the behavior of $V(t)$. Charging profiles for $T = 300\,\text{K}$ and for $T = 296\,\text{K}$ are also shown (Reprinted from [127], with permission from Elsevier)

profiles for $T = 300\,\text{K}$ and for $T = 296\,\text{K}$ are also shown in order to hint as to how Eq. (3.16) behaves with temperature.

As can be easily noticed, the charging profiles for $T = 298\,\text{K}$ in Fig. 3.4 and Fig. 3.5 are exactly the same of those in Fig. 2.2 and Fig. 2.3 of Chapter 2. All this has a twofold meaning. First of all, even with η_{ext} five orders of magnitude greater than η_{in}, the counter-emission disturbance on the charging process is negligible and, as mentioned before, it would become even more negligible with the suggested choice of $\overline{\eta}_{ext} \ll \overline{\eta}_{in}$. More importantly, the preliminary model of the physical process introduced in Chapter 2 is definitely robust against the approximation made there, namely $\phi_{ext} = \infty$ or $\phi_{ext} \gg \phi_{in}$. Moreover, from the experimental point of view the physical process appears to be not so sensitive to the specific value of ϕ_{ext}, provided that $\phi_{ext} > \phi_{in} \sim 1\,\text{eV}$.

3.2 Discussion

In this section we shall investigate the possibility of a measurable current flowing across a thermo-charged capacitor once its plates are electrically shorted through a suitable resistor. Our main goal is to show why the generation of potentially usable voltage, current and electric power out of a single thermal source at room temperature (as if it were an ordinary battery) appears not to be forbidden by physics. We also try to answer some objections, which should naturally arise against the results reached so far.

As explained in Section 3.1, thermionic emission produces a bias between the plates of the spherical capacitor (Figs. 3.4 and 3.5). A similar bias also originates across the metal/semiconductor junction (since the Ag–O–Cs layer charges up positively) of the inner sphere: given the rectifying nature of the metal/Ag–O–Cs junction, the bias has the characteristics of a *reverse bias*. As described in Section 3.1, the reverse bias causes a *reverse leakage density current* j_0 which slowly transfers electrons from the inner metallic sphere to the positively charged Ag–O–Cs layer. This reverse leakage density current j_0 is microscopic; for typical metal/semiconductor junctions it usually ranges from 10^{-6} to $10^{-9}\,\mathrm{A/cm^2}$ with reverse biases of V or 0.1 V. Its intensity is weakly dependent on the magnitude of the reverse bias, provided that such a bias is below the breakdown voltage of the junction [25, 44, 76, 86]. As seen in the previous analysis, the bias across the thermo-charged capacitor is far below this voltage.

Now, the mere fact that a non-zero, almost constant, reverse bias leakage density current j_0 exists, although tiny, allows the thermo-charging process to be potentially exploitable. As a matter of fact, if we suitably increase the surface area of the inner sphere S_a (and that of the outer one, accordingly), then the contact area between the Ag–O–Cs layer and the metal increases. This means that, in principle, it is possible to obtain a macroscopic reverse leakage current (RLC) that allows the rather quick transfer of the voltage drop to both terminal leads of the capacitor, since RLC is equal to $j_0 S_a$.

Let us imagine, for the sake of argument, to build a room-sized thermo-charged capacitor with radii $a = 100\,\mathrm{cm}$ and $b = 200\,\mathrm{cm}$. In this case, the inner sphere surface is equal to $S_a = 4\pi a^2 \approx 10^5\,\mathrm{cm^2}$. Thus, the total reverse leakage current $j_0 S_a$ should vary between 0.1–$100\,\mathrm{mA}$; this is a quite macroscopic current. The thermo-charging profile for the capacitor with the following parameters $\phi_{in} = 0.7\,\mathrm{eV}$, $\phi_{ext} = 4.0\,\mathrm{eV}$, $b = 2.0\,\mathrm{m}$, $T = 298\,\mathrm{K}$, $\bar{\eta}_{in} = 10^{-5}$ and

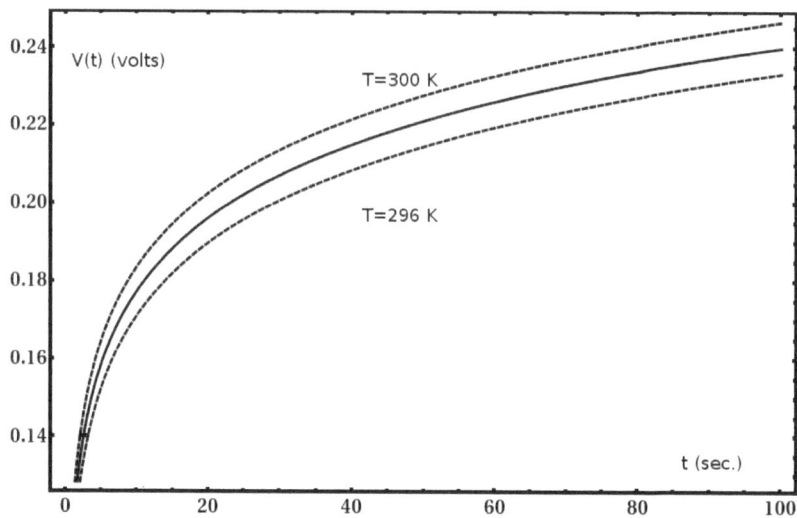

Figure 3.6: Thermo-charging profile for the spherical capacitor described in Section 3.2 with $\phi_{in} = 0.7\,\text{eV}$, $\phi_{ext} = 4.0\,\text{eV}$, $b = 2.0\,\text{m}$, $T = 298\,\text{K}$, $\overline{\eta}_{in} = 10^{-5}$ and $\overline{\eta}_{ext} = 1$. Charging profiles for $T = 300\,\text{K}$ and for $T = 296\,\text{K}$ are also shown (Reprinted from [127], with permission from Elsevier)

$\overline{\eta}_{ext} = 1$, is represented in Fig. 3.6.

Let us now compare RLC in this case with the thermionic current across the vacuum region between the spheres $i_{ti} = dQ/dt$ at $t = 0\,\text{s}$ (and $V(0) = 0\,\text{V}$). It should be clear that the value of i_{ti} at $t = 0\,\text{s}$ is the maximum value reachable by the thermionic current during the charging process. By rearranging Eq. (3.12) and Eq. (3.16) we obtain,

$$i_{ti}(t) = \frac{dQ}{dt} = \frac{2\pi^2 eb^2}{c^2}\left(\frac{kT}{h}\right)^3\left(\overline{\eta}_{in}\int_{\frac{eV(t)+\phi_{in}}{kT}}^{\infty}\frac{x^2 dx}{e^x - 1} + \right.$$
$$\left. - 4\overline{\eta}_{ext}\int_{\frac{\phi_{ext}}{kT}}^{\infty}\frac{x^2 dx}{e^x - 1}\right), \quad (3.18)$$

and through numerical calculations for $t = 0\,\text{s}$ we get $i_{ti}(0) \approx 3.91 \times 10^{-10}\,\text{A}$.

We note that $i_{ti} \ll j_0 S_a$, meaning that the voltage drop thermally gained within the plates of the capacitor is quickly transferred to both terminal leads of the capacitor and can be directly detected through an electrometer.

In general, we can directly compare the reverse bias leakage density current j_0 to the thermionic density current, $j_{ti}(0)$, obtained from Eq. (3.18) as follows,

$$j_{ti}(0) = \frac{i_{ti}(0)}{S_a} = \frac{i_{ti}(0)}{4\pi a^2} \approx 3.11 \times 10^{-15} \, \text{A/cm}^2, \qquad (3.19)$$

by assuming the maximizing condition $a = b/2$ for dV/dt (see Eq. 3.15). Thus, for $\phi_{in} = 0.7 \, \text{eV}$, $\phi_{ext} = 4.0 \, \text{eV}$, $T = 298 \, \text{K}$, $\overline{\eta}_{in} = 10^{-5}$, $\overline{\eta}_{ext} = 1$ and j_0 in the reasonable range given before, we see from Eq. (3.19) that j_0 is *always* greater that $j_{ti}(0)$: this roughly means that the voltage drop thermally gained within the plates of the capacitor is *always* quickly transferred to both terminal leads of the capacitor, no matter how big or small the capacitor. Obviously, the larger is the capacitor, the greater the current "flowing" through it.

Let us now consider a thermo-charged capacitor shorted through a suitable resistor R. In this case, the capacitor behaves like a battery that dissipates its power through the resistor (i.e. the Joule effect). Consider the electrical circuit depicted in Fig. 3.7. In steady-state conditions, voltage drop V_s and current i_s should be the same across the capacitor and the resistor; moreover, they should be constant in time. According to Ohm's Law, we must have $R = V_s/i_s$; this relation also gives the numerical value of the resistance R to be used in order to have these particular values of V_s and i_s. Note that this resistance turns out to be of the order of Tera-ohms (TΩ): this gives an indication of how great the input impedance of a measuring electrometer must be in order to be able to detect anything (if it exists at all). Given the equations of the thermo-charged capacitor described before, we must have,

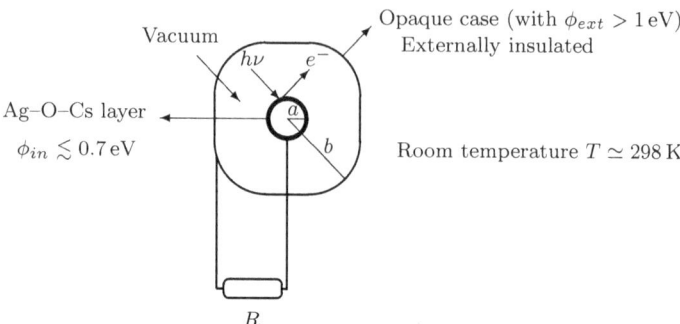

Figure 3.7: Thermo-charged spherical capacitor shorted through a resistor R (Reprinted from [127], with permission from Elsevier)

$$i_s = \frac{2\pi^2 e b^2}{c^2}\left(\frac{kT}{h}\right)^3\left(\overline{\eta}_{in}\int_{\frac{eV_s+\phi_{in}}{kT}}^{\infty}\frac{x^2 dx}{e^x-1} - 4\overline{\eta}_{ext}\int_{\frac{\phi_{ext}}{kT}}^{\infty}\frac{x^2 dx}{e^x-1}\right).$$
$$(3.20)$$

The electric power output P_s of the thermo-charged capacitor is then calculated as,

$$P_s = V_s i_s = \frac{2\pi^2 e b^2 V_s}{c^2}\left(\frac{kT}{h}\right)^3\left(\overline{\eta}_{in}\int_{\frac{eV_s+\phi_{in}}{kT}}^{\infty}\frac{x^2 dx}{e^x-1} + \right.$$
$$\left. - 4\overline{\eta}_{ext}\int_{\frac{\phi_{ext}}{kT}}^{\infty}\frac{x^2 dx}{e^x-1}\right). \quad (3.21)$$

The power output per unit surface of the inner sphere $\mathcal{P}_s$ (specific power output) is thus,

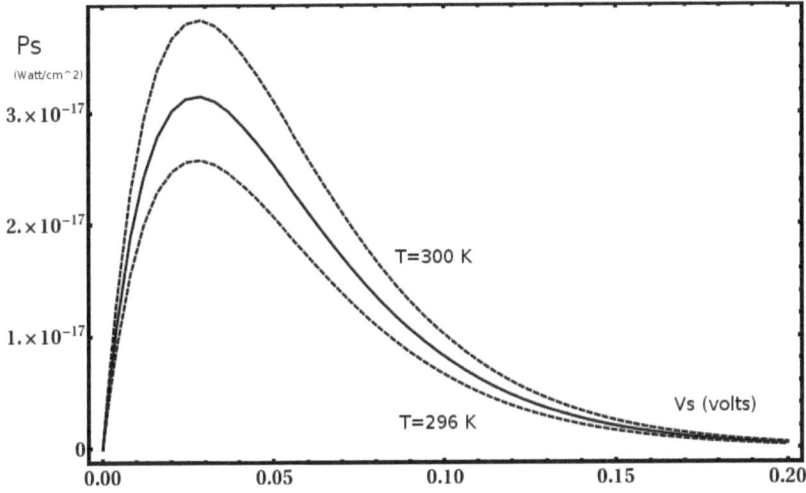

Figure 3.8: Power output per unit surface area of the inner sphere, $\mathcal{P}_s$, against voltage drop V_s of the electrical circuit capacitor/resistor depicted in Fig. 3.7. Power outputs for $T = 300\,\text{K}$ and for $T = 296\,\text{K}$ are also shown (Reprinted from [127], with permission from Elsevier)

$$\mathcal{P}_s = \frac{V_s i_s}{S_a} = \frac{2\pi e V_s}{c^2}\left(\frac{kT}{h}\right)^3 \left(\overline{\eta}_{in}\int_{\frac{eV_s+\phi_{in}}{kT}}^{\infty}\frac{x^2 dx}{e^x - 1} + \right.$$
$$\left. - 4\overline{\eta}_{ext}\int_{\frac{\phi_{ext}}{kT}}^{\infty}\frac{x^2 dx}{e^x - 1}\right). \quad (3.22)$$

In Fig. 3.8, the specific power output $\mathcal{P}_s$ is plotted against the steady-state voltage V_s. For the capacitors described in this section, namely that with $a = 10\,\text{cm}$, $S_a \approx 1260\,\text{cm}^2$ and that with $a = 100\,\text{cm}$, $S_a \approx 10^5\,\text{cm}^2$, we obtain $P_{max} \approx 4 \times 10^{-14}\,\text{W}$ and $P_{max} \approx 4 \times 10^{-12}\,\text{W}$, respectively. These are quite minute power outputs indeed, also considering that the second capacitor has "uncomfortable" room-sized dimensions.

There are few doubts now that one of the definite results of the entire analysis we have made so far is that thermo-charged capacitors are not particularly efficient devices. Even so, it is important to test their functioning experimentally, since if they work according to the analysis carried out thus far, we would have a *reproducible* second law violation. We believe that the glaring *smallness* of this violation is a secondary problem that can be overcome with further research, provided that this *smallness* does not prevent a clear and unambiguous result from being singled out from other environmental disturbances.

We have just seen that current and power output of circuits like those depicted in Fig. 3.7 have orders of magnitude of 10^{-10} – 10^{-14}, and even smaller if the physical dimensions of the spheres are centimetric. Nevertheless, the voltage drop of non-shorted capacitors is macroscopic already for a single capacitor: it is of the order of 0.1 V. Thus, it should not be difficult to build 10s of cm-sized vacuum capacitors, wired in series, so as to produce a voltage drop of V or 10s V, provided that an extremely sensitive, high-input impedance (of the order of $T\Omega$) electrometer is used as measuring apparatus (e.g. Keithley 6514 electrometer). We will deal with the measurement issue in Chapter 5.

From an experimental perspective, some manufacturing difficulties have to be faced. For instance, the stable ultra-high vacuum required inside the capacitor may pose a technical challenge, but we believe it can be so only for m-sized devices. Vacuum technology is currently quite sophisticated and mature: we expect that the construction of light, vacuum-proof cm-sized capacitors should not pose any concern at all. Moreover, some possible thermoelectric disturbances, like thermocouple and Thomson effects[3], can be reduced or canceled out through the proper design of the devices [127].

One may wonder why a voltage drop due to the thermo-charging process seems never to have been observed so far within vacuum tubes. After all, cm vacuum tubes have been widely used in electronic devices (phototubes, photomultipliers, radios, TVs, etc.) for a long time before the discovery of silicon (photo)diodes. One possible answer is that the thermo-charged capacitor is a ultra-

---

[3]Voltage/current generation due to temperature gradient within a system made of a single material.

high impedance source (TΩ to 10s TΩ) and its output cannot be detected at all with cheap commercial electrometers; its output is probably near the theoretical limits of voltage measurement [107]. In order to succeed, one needs expensive ultra-high input impedance electrometers (with reduced loading errors) and the awareness of the minuteness of what you are going to measure. The effect described here is a really tiny one (power output $\approx 10^{-14}$ W) and may be easily masked by a voltage offset due to the electrometer input bias current[4] during direct measurements. Or, if a signal is actually detected, it may be confused with other known thermoelectric effects (thermocouple/Thomson effects). Lastly, commercial vacuum (photo)tubes are obviously not thought or designed to magnify a possible thermo-generated output.

Finally, let us anticipate (and try to exhaustively respond to) some possible criticism. We shall address some more subtle issues in the next chapter. Among the main objections to our results, probably the first one is that our device appears to suffer the same shortcomings of the self-rectifying diode scheme (see, for example, [15] and [27] and references therein). We have to stress that the thermo-charged capacitor is a quite different device from a solid-state diode: in solid-state diodes, the two terminals are actually in physical contact with one another inside the diode through the n-p junction, and the dynamic balance between the built-in electric field and the diffusion forces across the junction prevent the establishment of a non-random charge displacement far from the depletion region toward the terminals. Thus, this prevents the creation of a voltage drop between the terminal leads.

In thermo-charged capacitors, on the contrary, the presence of a vacuum between the inner and the outer spheres (there is no physical contact between them inside the capacitor), and the fact that only one sphere is covered with a low work function material, appear to exclude the dynamic balance above, and a definite, one-way migration of charges across the vacuum region between

---

[4]The input bias current flows at the instrument input due to internal instrument circuitry and the internal bias voltage.

the plates (to the terminal leads) should be possible, as described above. On this basis, we will go into more detail in the next chapter. The space charge that settles between the plates (and that could interfere with the expected outcome) is also handled in some way.

3.3   Electro-mechanical analog of a thermo-charged capacitor

The thermo-charged capacitor has an easily understandable electro-mechanical analog, which is both instructive and explicative and thus worthwhile to describe.

Consider the device depicted in Fig. 3.9-1. It is essentially a parallel-plate capacitor with one plate made of a metal with relatively low work function, e.g. zinc (Zn), and the other plate made of a metal with relatively high work function, e.g. copper (Cu). This last plate is also free to move in space. Moreover, a copper wire (with a load R) is connected to the zinc plate through a small junction Cu-Zn. The Cu-Zn junction generates a very thin[5] depletion layer along the contact surface, where positive and negative charges are locally displaced after the dynamically balanced drift of electrons from zinc to copper through the contact area (Fig. 3.9-1).

The first step of the electro-mechanical cycle we are going to describe consists of moving the copper plate toward the zinc one until the contact. In this phase, no significant external work is required. After the contact, a second (and larger) depletion layer forms between zinc and copper plates. Also in this case, the new Cu-Zn junction generates a very thin depletion layer along the new (and wider) contact surface (Fig. 3.9-2).

In step two, an external work L is applied to the copper plate in order to remove it from the zinc plate: L is different from zero now, since the charge displacement across the new Cu-Zn junction makes the two plates attract each other. When the two plates are again suitably removed, the charges initially localized within the

---

[5]The junction being a metal to metal one.

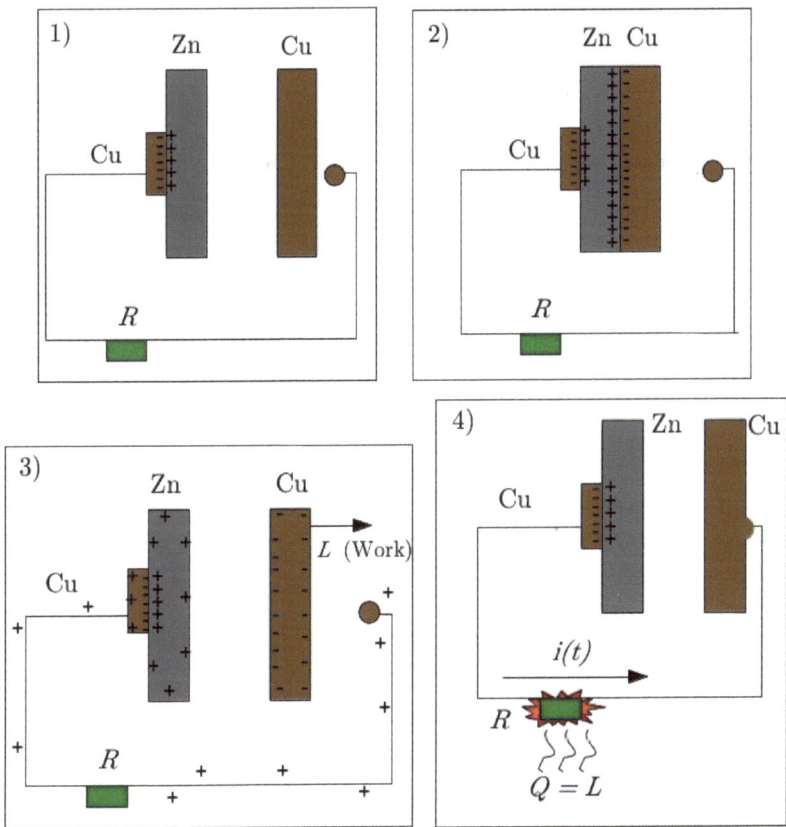

Figure 3.9: Electro-mechanical analog of a thermo-charged capacitor (Reprinted from [127], with permission from Elsevier)

thin depletion layer are free to spread across the surfaces of the two metallic plates and wire, satisfying equi-potentiality (Fig. 3.9-3; see, for example, [48]).

In the third step of the cycle, we put the negatively charged copper plate and the positively charged copper wire into contact (Fig. 3.9-4). Just after the contact, electrons start to flow from the copper plate to the zinc plate across the copper wire/load until both plates become neutral. Due to the Joule effect during the discharge, the load R heats up. The amount of heat Q released to the environment is nearly equal to the external work L done to the system in step number two (according to the first law of thermodynamics).

The fourth step, which closes the cycle, is trivial and consists of returning the copper plate to its initial position, as in Fig. 3.9-1.

The analogy between this scheme and the functioning of the thermo-charged capacitor should be clear: both devices work with materials having different work functions, and in both devices current (electrons) flows across the contact junctions so as to reestablish electrical equilibrium. The main difference is in the source of the energy that sustains the current flow. In the electro-mechanical scheme, the source is the external work L done to the system through the movable copper plate. In the thermo-charged capacitor, it is black-body radiation (of the uniformly heated environment) that yields kinetic energy to the electrons and lets them fly out into the external (fixed) plate of the capacitor.

Chapter 4

Theoretical corroboration

In this chapter, we shall address two delicate issues that have not been tackled explicitly enough in the previous chapters. They are actually crucial for the theoretical tenability of our main contention of thermo-charged capacitor as a second law violator. Once more, we have to apologize to the careful reader: confronted with the decision to favor clarity and self-containedness over brevity, we have chosen to repeat some of the preliminaries made in the previous chapters (actually, we shall go more into detail here). May the reader excuse this apparent waste of his/her time by recalling the old Latin saying: *repetita iuvant*. As a matter of fact, the results achieved in this chapter provide a further theoretical corroboration of the theory behind the thermo-charged capacitor. At the end of the chapter, a conventional photoelectric emission experiment is also proposed to test thermo-charged capacitor functioning indirectly.

In Figure 4.1, a further sketch of a thermo-charged capacitor is shown, along with the features essential to the following treatment. With respect to similar schematics given in previous chapters (spherical thermo-charged capacitors in Fig. 2.1 and in Fig. 3.1), here we turn to a simpler structural design (flat capacitor) because our focus is to investigate theoretically what happens across the contact surfaces $(1-2)$ and between all the free surfaces $(2-3$ and $4-5)$.

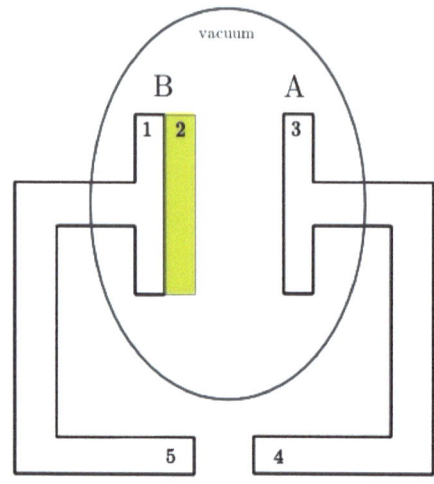

Figure 4.1: Sketch of a (flat) vacuum thermo-charged capacitor: electrode A $(3-4)$ is made of a high work function metal, electrode B $(1-5)$ is made of the same high work function metal coated with a layer (2) of semiconductor Ag–O–Cs with low work function. Work functions are such that $\phi_{1\to5} = \phi_{3\to4} > \phi_2$.

Electrode A is made of a metallic material with relatively high work function ($\phi_A = \phi_{3\to4} > 1\,\mathrm{eV}$). Electrode B is made of the same conductive material as A, but is coated with a layer (2) of semiconductor Ag–O–Cs ($\phi_2 \lesssim 0.7\,\mathrm{eV}$ and $\phi_{1\to5} = \phi_A$).

The aim of this chapter is to answer satisfactorily the legitimate doubt that even if the mathematical treatment made in the previous chapters was unexceptionable, it might be incomplete: we might have missed some subtle physical processes that prevent the formation of a potential drop between the capacitor terminals 4 and 5.

Since the publication of the first papers on the thermo-charged capacitor, no one seems to have put forward any criticism. This notwithstanding, we have devoted some time and energy to address at least two possible objections concerning the design and functioning. The first is related to the presence of a rectifying Schottky junction inside electrode B (between the metal and the Ag–O–Cs layer), which appears to prevent any electron displacement from the metal to the semiconductor, and thus precluding any current flow across the whole capacitor when terminals 4 and 5 are shorted (or, equivalently, giving zero voltage drop between 4 and 5 when the circuit is left open). Actually, this objection was addressed and settled beyond reasonable doubt in Chapter 3.

The second objection has been the topic of an addendum to the second paper on thermo-charged capacitors [128]. Here we shall resume its key aspects.

As already noted, the contact surface between the metallic part 1 of electrode B and its semiconductor layer 2 (Ag–O–Cs) is a metal/n-type semiconductor junction (Fig. 4.1, region $1-2$). Across such a junction, a contact potential builds up, equal to the difference between the two work functions divided by the electronic charge, $\Delta V = \frac{\phi_1 - \phi_2}{e}$. This potential is the result of charge diffusion across the junction $1-2$ as soon as the two materials are physically joined. The junction is the region where, at equilibrium, a balance between electrostatic and diffusive (thermally driven) forces is attained. Electrostatic forces tend to reestablish electrical neutrality (by pulling the electrons back) against diffusive forces.

It is a widely held belief that a voltage drop builds up not only across the contact surface $1-2$, but also instantaneously between the surfaces at the free ends of the joined materials (free surface of semiconductor 2 on one side and free surface 5 of the metal on the other, but also free surface of semiconductor 2 and free surface 3 of electrode A, when terminals 4 and 5 are shorted, see Fig 4.1).

Note that this voltage drop is not intended to be that generated by thermionic emission of electrons from 2 to 3. It is intended to originate from an overall charge displacement in the bulk of electrode B across the junction $1-2$ as soon as the materials 1 and 2 are physically joined, and is said to be of the same magnitude of the contact potential ΔV. To our knowledge, no textbook or published paper on the subject clearly explains how and why these charges collectively and macroscopically move inside the bulk of electrode B across the junction so as to charge the metal $(1 - 5$, or $1 - 5 - 4 - 3$ when terminals 4 and 5 are shorted) negatively and semiconductor 2 positively. As already noticed, all this is usually explained by appealing to a supposedly straightforward application of the Kirchhoff's second (loop) rule.

If this were true, it would prevent the formation and the flow of any net thermionic current across the whole capacitor when the two electrodes are electrically shorted at their free ends (terminals 4 and 5), so as to establish a closed circuit. As a matter of fact, in order to reach electrode A, any thermionic electron escaping electrode B would need the same energy needed by an electron escaping electrode A to reach electrode B. B-electrons need an energy equal to $\phi_2 + e\Delta V$, since they must be ejected (requiring energy ϕ_2) and then have to overcome the alleged potential drop ΔV instantaneously generated between 2 and 3 owing to the contact between 1 and 2 (energy equal to $e\Delta V$). A-electrons need ϕ_3 (only the energy to be ejected). However, since $e\Delta V = \phi_1 - \phi_2 = \phi_3 - \phi_2$, the B-electrons need $\phi_2 + e\Delta V = \phi_3$; the same energy required by electrons from electrode A.

We have explicitly shown in Chapter 3 that no voltage drop, and thus no electric field, builds up between the surfaces at the free ends of two materials with different work functions (namely, between 2 and 5 or, equivalently, between 2 and 3 when terminals 4 and 5 are shorted) when the materials are physically joined at one end (region $1 - 2$).

Here, though, we present a similar proof of this tenet, which is more akin to that originally published in [128, 134] and that is, as we shall see, necessary for the analysis we will carry out in the next section. We perform a sort of "closed-loop" analysis; namely we make an explicit application of the *path-independence* law and/or Kirchhoff's loop rule. The physical principle that forms the basis of these two laws is the more fundamental law of conservation of energy.

The schematic in Fig. 4.2 equivalently represents either the sole electrode B of Fig. 4.1 or, if terminals 4 and 5 are shorted, the whole thermo-charged capacitor of Fig. 4.1. J-I represents always

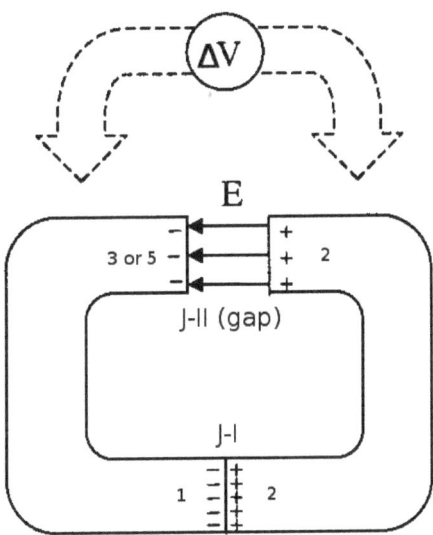

Figure 4.2: This figure equivalently represents either the sole electrode B of Fig. 4.1 or, if terminals 4 and 5 are shorted, the whole thermo-charged capacitor of Fig. 4.1. J-I is the physical junction $1 - 2$ and J-II is the gap between 2 and 5, if this figure is intended to be the sole electrode B, or the gap between 2 and 3 if this figure is intended to be the whole thermo-charged of Fig. 4.1 with terminals 4 and 5 shorted. Work functions are such that $\phi_1 > \phi_2$. (Adapted from [95])

the junction $1 - 2$, while gap J-II represents the gap between 2 and 5 of Fig. 4.1, if Fig. 4.2 is intended to be the sole electrode B, or the gap between 2 and 3 if Fig. 4.2 is intended to be the whole thermo-charged of Fig. 4.1 with terminals 4 and 5 shorted. At equilibrium, conservation of energy demands that a test electronic charge e conveyed around a closed path γ in the device bulk of Fig. 4.2, through physical junction J-I and across gap J-II, must undergo zero net work from *all* the forces present along the path. Mathematically, we must have,

$$\oint_\gamma dW_{tot} = 0. \tag{4.1}$$

Note that, for now, we ignore the thermionic emission of all the materials. For this analysis it is as though thermionic emission between 2 and 3 were turned off. We only take into account the physical process across the contact junction $1 - 2$ at equilibrium.

At equilibrium, the only two regions where forces are allowed to be non-null are the J-I and J-II regions. An electric field elsewhere in the device bulk (other than in the contact region) would generate a current, which contradicts the assumption of equilibrium. When the test charge e crosses J-I, it is subject to the built-in electric field force $e\mathbf{E}_{bi}$ (related to the junction contact potential ΔV as $\Delta V = x_d \mathbf{E}_{bi}$, where x_d is the depletion region width) and to the diffusion force $\mathbf{F}_{diff}$. This latter "force" is the thermally driven force responsible for the establishment of the contact potential at J-I. We know that at equilibrium $e\mathbf{E}_{bi} = -\mathbf{F}_{diff}$ and that $\mathbf{F}_{diff}$ is different from zero and constantly present; otherwise $\mathbf{E}_{bi}$ (and ΔV) would soon drop to zero. Thus:

$$0 = \oint_\gamma dW_{tot} = \int_{\text{J-I}} (e\mathbf{E}_{bi} + \mathbf{F}_{diff}) \cdot d\vec{\gamma} + \int_{\text{J-II}} dW_{ext} = 0 + \int_{\text{J-II}} dW_{ext}. \tag{4.2}$$

In the J-II gap there are no diffusion forces, since it is a vacuum gap; eventually we have,

$$0 = \int_{\text{J-II}} dW_{ext} = \int_{\text{J-II}} e\mathbf{E}_{\text{J-II}} \cdot d\vec{\gamma} = e|\mathbf{E}_{\text{J-II}}|x_g \quad \rightarrow \quad |\mathbf{E}_{\text{J-II}}| = 0, \tag{4.3}$$

where x_g is the gap width. Zero electric field means no voltage drop.

In the following section, we perform the same "closed-loop" analysis across the whole thermo-charged capacitor after having, this time, "switched on" the thermionic emission between 2 and

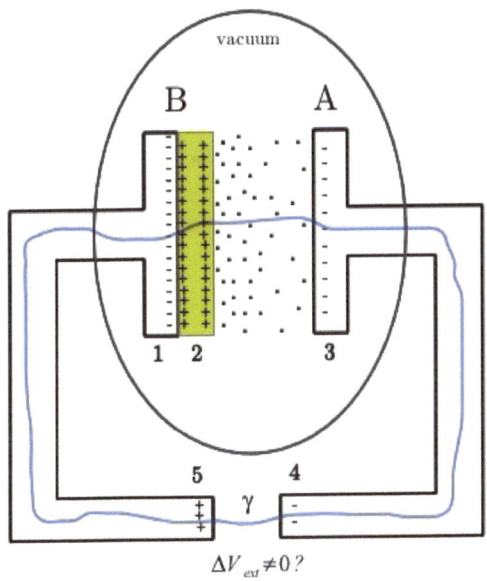

Figure 4.3: Charged capacitor at equilibrium with ambient heat (uniform temperature) with non-shorted terminal leads 4 and 5. As in Fig. 4.1, region $1 - 2$ is the metal/Ag–O–Cs junction. Dots in the vacuum region (region $2 - 3$) are the space charge electrons, which, at equilibrium, are continually emitted and re-absorbed by the electrodes' surfaces. γ is the closed path traveled by the test charge e.

3, and thus by considering the capacitor as thermionically charged and at equilibrium (with terminals 4 and 5 not shorted) [138].

4.1 Closed-loop analysis across a thermo-charged capacitor

We apply the energy conservation analysis made in the previous section to the whole capacitor, now thermionically charged and at equilibrium. In what follows, reference is made to Figure 4.3.

Figure 4.3 shows a thermionically charged capacitor at equilibrium with non-shorted terminal leads. As in Fig. 4.1, region $1 - 2$ is the metal/Ag–O–Cs junction. The dots in the vacuum region (region $2 - 3$) represent the space charge electrons, which, at equilibrium, are continually emitted and re-absorbed by the electrodes' surfaces (the main part of these electrons comes from the Ag–O–Cs layer on electrode B). Part of the electrons emitted by electrode B has been definitely absorbed by electrode A and is represented by the minus-signs on it.

Once again, conservation of energy demands that a test electronic charge e conveyed around a closed path γ in the device bulk of Fig. 4.3 at equilibrium, through region $1 - 2$ (physical junction) and across regions $2 - 3$ (vacuum gap) and $4 - 5$ (open terminal leads), must undergo zero net work from *all* the forces present along the path. At equilibrium, the only regions where the forces are allowed to be non-zero are $1 - 2$, $2 - 3$ and $4 - 5$; mathematically, we have,

$$0 = \oint_{\gamma} d\mathrm{W}_{tot} = \int_{1-2} d\mathrm{W} + \int_{2-3} d\mathrm{W} + \int_{4-5} d\mathrm{W}. \qquad (4.4)$$

The integral $\int_{4-5} d\mathrm{W}$ is equal to $e\Delta V_{ext}$; namely the voltage drop at the free ends of the thermo-charged capacitor multiplied by the test charge e. The integral $\int_{1-2} d\mathrm{W}$ has already been proven to be equal to 0 (see Eq. 4.2).

The question is therefore: is the integral $\int_{2-3} d\mathrm{W}$ different from zero? We have already shown in the previous chapters that a voltage drop ΔV should arise inside the vacuum capacitor owing to the thermo-charging process, and thus an electric field $\mathbf{E}_{int}$ (nearly equal to $\frac{\Delta V}{d}$, where d is the inter-plate distance) should be present between the plates inside the capacitor (i.e. between 2 and 3).

Nevertheless, are we sure that a sort of compensating (thermally driven) diffusion force $\mathbf{F}_{int}$, similar to that present in the contact junction $1 - 2$, is not present inside the vacuum capacitor that cancels out the internal electric force $e\mathbf{E}_{int}$? If this were the case, we would have,

$$-e\mathbf{E}_{int} = \mathbf{F}_{int} \tag{4.5}$$

and thus,

$$0 = \oint_\gamma dW_{tot} = 0 + \int_{2-3} dW + \int_{4-5} dW =$$

$$= \int_{2-3} (e\mathbf{E}_{int} + \mathbf{F}_{int}) \cdot d\vec{\gamma} + e\Delta V_{ext} =$$

$$= 0 + e\Delta V_{ext}. \tag{4.6}$$

This would mean that $\Delta V_{ext} = 0$: the thermo-charged capacitor would have a zero voltage drop between its external leads.

Let us go into the possible nature of $\mathbf{F}_{int}$. As in the case of contact junction, this force could be seen as the collective and macroscopic manifestation of single microscopic actions on the electrons ejected by the thermionic surfaces (mainly, the Ag–O–Cs layer). At equilibrium, electrons are continuously emitted (due to the absorption of light quanta from black body radiation) and re-adsorbed by the surface of electrode B (the same process also takes place on electrode A but at a negligible rate). The collective action of quanta absorption could be seen as a force acting in the opposite direction of $\mathbf{E}_{int}$: the field tends to pull the electrons just ejected (and thus also the test charge e) from electrode B back to electrode B. The force $\mathbf{F}_{int}$ tends instead to push electrons (and thus the test charge e) away from electrode B.

In the following subsection, we put forward three arguments that suggest that $\mathbf{F}_{int}$ is actually a misleading concept.

4.1.1   Discussion

We now list three arguments, in increasing order of cogency, which appear to dismiss any concern about the possibility that $\mathbf{F}_{int}$ really exists and cancels out the internal, thermally generated, field force $\mathbf{E}_{int}$:

a) Cursory objection to $\mathbf{F}_{int}$: contrary to what happens in the junction depletion region, inside the thermo-charged capacitor there are no diffusion forces ($\mathbf{F}_{int} = 0$) because there is a vacuum between electrodes A and B.

b) Heuristic objection to $\mathbf{F}_{int}$: before, when talking about contact junctions, we emphasized that the diffusion "force" $\mathbf{F}_{diff}$ acting upon the electrons must be continuously present in the depletion region; otherwise the built-in electric field $\mathbf{E}_{bi}$ would go instantaneously to zero (there is a physical contact between metal and semiconductor in the junction). This close connection between $\mathbf{F}_{diff}$ and $\mathbf{E}_{bi}$ is expressed by the identity $\mathbf{F}_{diff} = -e\mathbf{E}_{bi}$. On the other hand, in a fully charged thermo-charged capacitor at equilibrium, if one could "switch-off" thermionic emission, then the field inside the capacitor would still be there (maybe only decreased slightly and becoming more uniform since space charge would cease), because some electrons from electrode B have already been absorbed/collected by electrode A. Thus, we may heuristically conclude that, at equilibrium, $\mathbf{E}_{int}$ and $\mathbf{F}_{int}$ do not depend upon each other[1] and there would be no reason to assume that $\mathbf{F}_{int} = -e\mathbf{E}_{int}$ exactly. In this case, there is no strict cause/effect relation between $\mathbf{E}_{int}$ and $\mathbf{F}_{int}$ as in the case of a contact junction.

c) Comparison with the behavior of electrons in a photoelectric tube: given the possible microscopic explanation/origin of the force $\mathbf{F}_{int}$, this force would also be present on the active surfaces of a photoelectric tube[2] with electrodes submersed in and illuminated by diffused light and with non-shorted terminals. This time, photoelectric emission comes into play. If we apply the same closed-loop analysis performed on the thermo-charged capacitor and admit the equivalence $\mathbf{F}_{int} = -e\mathbf{E}_{int}$, then we would have a zero voltage drop between

---

[1]One does not exist *only* because of the other.
[2]The device depicted in Fig. 4.1 or Fig 4.3 could equally work as a photoelectric tube when illuminated by visible light (see Section 4.2).

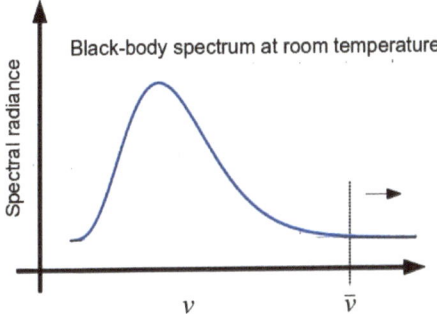

Figure 4.4: Any body in thermal equilibrium at any temperature T is surrounded by a bath of radiation in which the frequency distribution is given by Planck's formula. This formula places no finite limit on the magnitude of the frequencies occurring; so that there will always be a frequency $\bar{\nu}$ present for which $h\bar{\nu}$ is greater than the work function of the body and thus triggering electron emission.

the (open) external leads of an illuminated photoelectric cell. However, this is experimentally difficult to sustain.

4.2   Conventional photoelectric emission test

Here, we draw attention to an indirect but, from an experimental point of view, probably easier way to test the thermo-charging phenomenon. It is a further development of argument c) in the previous Subsection 4.1.1. We shall introduce a conventional photoelectric cell (UV spectrum) operating under highly symmetrical lighting and show why it could be useful to test thermo-charged capacitor functioning indirectly. The idea and the rationale behind the present proposal come from the following simple observation: if it is possible to extract energy from the plain (UV spectrum) photoelectric effect, why should it not be possible in principle to extract energy (although to a much lesser extent) from the photoelectric effect induced by the high-energy tail of the black-body spectrum (see Fig. 4.4) at a single temperature (provided that the device has two plates–anode and cathode–with different work functions)?

It could be objected that a UV spectrum photoelectric device works only when light impinges onto only one of the two plates, no matter whether they have different work functions; while in a "black-body spectrum" device, both plates are inevitably affected by the same radiation (the device is submersed in black-body radiation), no matter whether they have different work functions.

A relatively simple way to settle the issue (*experimentum crucis*) is to build a conventional photoelectric device (see Fig. 4.5) with equal parallel plates, same size but different work functions ($\phi_1 > \phi_2$), and to enlighten them both with the same amount of light, equal in frequency, intensity and geometry (for instance, by putting a point-like source of light right between the plates). If that photoelectric cell works with same UV light on both plates, it will also work with the high-frequency tail of the black-body spectrum on both plates. If the system becomes charged (one plate with positive and the other with negative charges due to excess electrons from plate ϕ_2), the same will happen, although to a much lesser extent, with "black-body spectrum" devices, and thus with thermo-charged capacitors.

This would mean an indirect but clear confirmation that thermo-charged capacitor behaves as expected. Moreover, this will also dissipate all the objections to the thermo-charged capacitor theory concerning rectifying contact junction and contact potential difference issues.

It must be admitted that the experimental test proposed here is not as simple as it appears at first sight. It is not simple to read its outcome (whatever it will be) as positive or negative. According to the theory, a negative result would be one in which no potential difference whatsoever is generated between the plates. However, it is practically impossible to obtain zero voltage in a real experiment; thus how is it possible to set the threshold between denial and confirmation? One possibility could be to turn the plates 180° around the central, point-like light source (left fixed) and see whether the voltage changes sign (or changes to a substantial degree) or stays almost the same. If there were a substantial change in the voltage output, this would mean that the output is exclusively ascribable to a residual asymmetry in lighting. If the voltage output remained almost the same, instead, there is room to believe that the photo-electric cell also works with the same UV lighting on both plates.

Obviously, a definitive answer to these questions can only come from a carefully designed laboratory test. This notwithstanding, given what we known about the photoelectric effect and considering the simplicity of the setup in Fig. 4.5, it is difficult to believe that the device remains uncharged upon symmetrical UV illumination and that this cycle cannot be repeated.

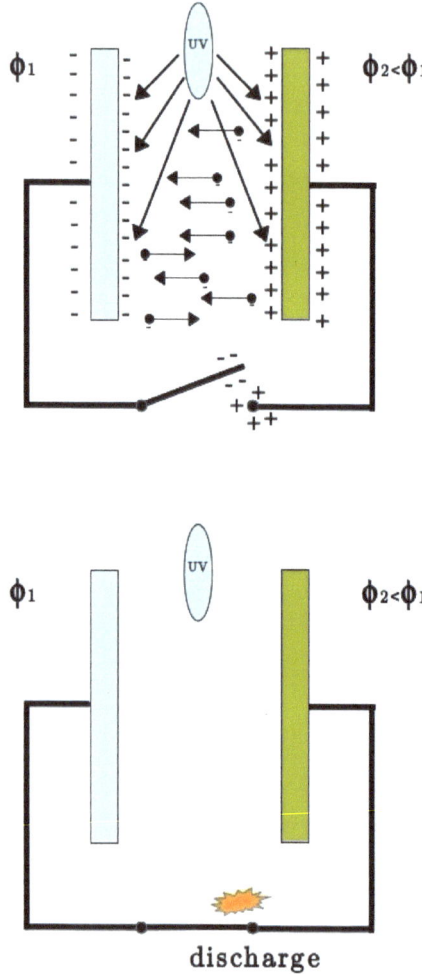

Figure 4.5: Schematic of a symmetrical photoelectric cell. Work functions ϕ_1 and ϕ_2 are such that $h\nu_{uv} > \phi_1 > \phi_2$ and thus photo-electrons from plate ϕ_1 are fewer than those from plate ϕ_2. If it works, this cycle can be repeated.

Chapter 5

Experimental corroboration

In recent years, two interesting experiments have been put forward, and actually carried out, allegedly instantiating a Maxwell's demon, i.e. a second law violator. In both experiments, the respective devices reportedly operate with thermionically emitted electrons bended by an external static magnetic field.

The first experiment was by Fu and Fu [136]. Their device is made of two similar plates A and B coated with photocathode Ag–O–Cs, placed side by side into a vacuum bulb at room temperature and electrically insulated from one another (see Figure 5.1). The two plates are geometrically similar (same dimensions) and their coatings have almost equal work functions, $\phi_A \simeq \phi_B$. The bulb (which is allegedly shielded from electromagnetic waves and external electric field by a copper box) is immersed in a static uniform magnetic field generated by a permanent magnet.

According to the authors, the external magnetic field plays the role of the demon, bending the otherwise symmetric thermionic electron flux preferably toward one of the electrodes (Figure 5.1) and hence generating a voltage drop, and a current when the bulb electrodes connected to the plates are externally shorted through a load. For an external load, they employed the internal resistance of

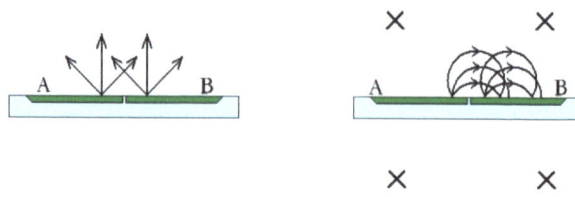

(a) Electrons ejected into a zero-field space   (b) Electrons ejected into a magnetic field

Figure 5.1: Sketched side view of the two similar electrodes inside the Fu and Fu vacuum bulb. [137]

the measuring apparatus, the high-input impedance electrometer Keithley 6514. The authors claim to have measured a maximum current of the order of 10^{-14} A and a maximum voltage drop of the order of 10^{-4} V. During the experiment, between one measurement and the other, the magnet was gradually moved toward the bulb from a relatively high distance ($\mathbf{B} \simeq \mathbf{0}$) in order to confirm the bending effect of the magnetic field: higher magnetic field means more bending and hence a larger current. During the magnet movement, however, the external magnetic field is non-static and thus some induced currents may arise. It is not clear from the paper whether the published data were recorded after a suitable relaxation time or after briefly short-circuiting the terminals of the bulb after every magnet movement. If we consider the circuit [bulb + electrometer internal resistance] as an RC circuit triggered by electromagnetic induction during the magnet movement, the RC time constant $\tau = RC$ can be estimated to be not as small: the electrometer's internal resistance is of the order of $R \sim 10^{10} - 10^{12} \, \Omega$, while the capacitance C of the circuit can be estimated to be not smaller than several μFs, thus $\tau \sim 10^4 - 10^6$ s. Thus, it is very difficult to relax the system in a reasonable time span.

The second experiment was by Perminov and Nikulov [130]. It bears interesting similarities to the experiment of Fu and Fu. Their device is sketched in Figure 5.2. Perminov and Nikulov make use of two electrodes with explicitly different work functions, separated by a dielectric (insulating) surface. The apparatus is put inside a

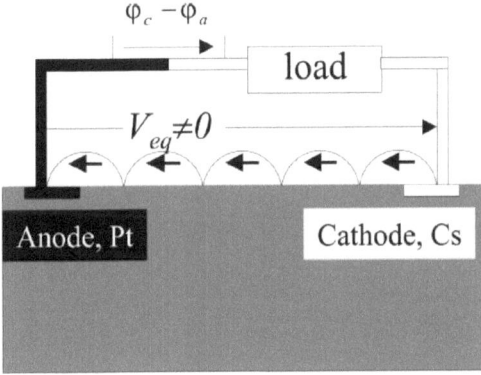

Figure 5.2: Working principle of the Perminov and Nikulov device [130]. The original figure caption in [130] reads: "The potential difference created because of the direct equilibrium movement of charged particles in magnetic field over the dielectric surface can exceed the contact potential difference between metals of the cathode and the anode, $V_{eq} > \phi_a - \phi_c$. Therefore a direct current in the electric circuit containing a load is possible under equilibrium conditions" (Reprinted from [130], with permission from AIP Publishing LLC)

vacuum chamber at a uniform temperature of about $100\,^\circ$C. Also in this case, there is an external static magnetic field, which is generated by a steady current flowing inside a linear conductor parallel to the dielectric surface. In this case, the magnetic field is supposed to act as a demon, pushing the thermionically emitted electrons from the low work function to the high work function electrode, and thus helping to overcome the electrodes' contact potential difference (see caption of Fig. 5.2). When they say that the magnetic field helps to overcome the electrodes' contact potential difference, they refer to the widely held belief, already discussed in the previous chapters, that when two materials with different work functions, ϕ_A and ϕ_B with $\phi_A > \phi_B$, are joined at one end[1], a voltage drop equal to $\Delta V = \frac{\phi_A - \phi_B}{e}$ builds up not only across the joined surfaces (where it is called the contact potential) but also instan-

---

[1]In this case and in that of the Fu and Fu device, this physical junction is somewhere along the circuit that connects the free electrode surfaces when the devices are shorted through a load.

taneously between the free ends of the materials (see Figure 3.3 in Chapter 3), where charges also accumulate, opposite to the voltage built up at the contact surface. Note that such a voltage drop is not intended to be that generated by thermionic emission in the vacuum gap. It allegedly forms earlier and instantaneously as a consequence of the physical junction at one end (J-I, Figure 3.3 in Chapter 3). This voltage drop would prevent any thermionically emitted electron in the gap from spontaneously moving from the low work function to the high work function free electrode surface (see Chapter 3).

As a matter of fact, even in the Fu and Fu experiment, the two electrodes cannot obviously have $\phi_A = \phi_B$ (and thus one must assume $\phi_A < \phi_B$ or $\phi_A > \phi_B$) because the Ag–O–Cs coatings are manufactured directly inside the bulb with cesium deposition on a silver layer and oxidation. Such a process cannot obviously guarantee exactly $\phi_A = \phi_B$. This means that even inside the Fu and Fu sample bulb, a voltage drop $\Delta V = \frac{\phi_B - \phi_A}{e}$ (or $\Delta V = \frac{\phi_A - \phi_B}{e}$) should build up between the free surfaces of the electrodes and represent a potential barrier to the thermionic electrons.

Thus, in both experiments, it seems that the external static magnetic field helps to push electrons across a (contact generated) potential barrier.

The experiment actually carried out by Perminov and Nikulov was not performed with the materials indicated in Figure 5.2. The authors used two equal tungsten electrodes (thus, $\phi_A \simeq \phi_B$) and claimed to have measured a current of the order of 10^{-7} A (across a circuit resistance of 1 MΩ). They also report that, according to their analysis, such a current is not ascribable to Thomson or Seebeck effects.

5.1   Discussion

We think that there are two weak points undermining both experiments, if not on the experimental/measurement side, at least from the theoretical modeling point of view. First, as explained more

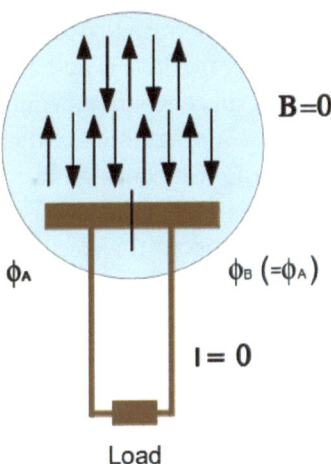

Figure 5.3: Case of $\phi_A = \phi_B$ in the Fu and Fu experiment. The arrows represent electrons in steady-state equilibrium inside the bulb (space charge). In order to break the equilibrium and generate a current, **B** is supposed to push the "flying" electrons from one electrode into the other, but this needs work; a static **B** cannot perform work (see Eq. 5.1) [137]

than once in this book, there is no macroscopic potential difference between the free ends of electrodes as instantaneous consequence of physically joining them at one end. Secondly, even if there actually were a "contact-generated" potential difference, a static uniform magnetic field cannot perform work. The Lorentz force $\mathbf{F} = -e\,\mathbf{v} \times \mathbf{B}$ ($-e$ is the electronic charge) cannot, by definition, perform any work, and thus it cannot be responsible for overcoming any mutual potential difference (contact-generated) between the free electrodes in both experiments. As a matter of fact, the infinitesimal work dW done by a static magnetic field $\mathbf{B}$ on a charge e moving with velocity $\mathbf{v}$ is given by,

$$dW = d\mathbf{x} \cdot \mathbf{F} = -e\,dt\,\mathbf{v} \cdot (\mathbf{v} \times \mathbf{B}) = 0 \qquad (5.1)$$

since for every vectors $\mathbf{a}$ and $\mathbf{c}$, we always have $\mathbf{a} \cdot (\mathbf{a} \times \mathbf{c}) = 0$.

Even if it were exactly $\phi_A = \phi_B$ (as Fu and Fu claim in their paper, but this also holds for the Perminov and Nikulov device on which tests were actually carried out), the static magnetic field would not be able to generate a voltage drop and a current between the terminals of the device. In the Fu and Fu experiment, for instance, the bulb is a closed space where an equilibrium space charge eventually settles. Without a magnetic field, electrons are emitted and absorbed by both Ag–O–Cs coatings in a sort of steady-state equilibrium (space charge is just that; see Figure 5.3). The glass envelope of the bulb contributes to the electrons' confinement. The static magnetic field can only modify the trajectory of each single electron in its movement back and forth between coatings and glass, but the overall equilibrium is preserved and the field cannot create a voltage drop and a current between the terminals of the sample: this would inevitably mean that the static magnetic field is able to do work, but this is physically impossible.

If we think of the steady-state (back and forth) electron movements as two equal and opposite (compensating) currents, it would be like asserting that a static magnetic field is able to produce the Hall effect inside a conductor where two equal and opposite currents (i.e. zero total current) flow.

We now outline two possible, mutually exclusive, scenarios to explain the experimental results described in the aforementioned papers. In the first scenario (I), the experimental results are not genuine and are the result of measurement/design flaws and/or environmental disturbances/interferences (natural and man-made electromagnetic waves, cosmic rays, or Thomson/Seebeck effects, etc.). In the second scenario (II), the experimental results are real. In this case, we appeal to the theoretical model discussed so far in this book.

Scenario I

The experimental results are not genuine and are the result of measurement/design errors and/or environmental disturbances/interferences. In this case, there is nothing really new to talk about. We think that it is actually difficult for people reading [130, 136] to evaluate possible procedural/measurement shortcomings. As a matter of fact, the experimental procedures followed by the authors of both experiments seem not to be completely "watertight" or, at the least in our opinion, the authors have not been completely successful in describing them as "watertight". For instance, we would have expected more simple cross-checks in both cases in order to exclude possible "external" sources of the measured voltage/current.

In the case of the Fu and Fu experiment, first of all, we would have tackled in some way the "relaxation time" issue described at the beginning of the chapter, which is not as trivial as it may appear. Secondly, we would have made a further test using a second bulb, identical to the sample one in all respect exception made for the Ag–O–Cs coating of the electrodes. If a non-zero current could be detected in this case, the origin of the signal would in both cases probably be in the interaction between the measurement apparatus (highly sensitive electrometer) and the external magnetic field.

In the Perminov and Nikulov experiment, the authors actually made some control measurements (zero magnetic field, inversion of the magnetic field[2]) obtaining apparently consistent results; but the high temperature of the test environment ($\sim 100\,^{\circ}\mathrm{C}$) and the high current used to generate the external magnetic field ($I_{\mathbf{B}} \simeq 600$ A) leave us cautious about the definitive exclusion that the results are actually a manifestation of interference with/disturbance to the measurement apparatus.

Scenario II

The experimental results are genuine. This second scenario is undoubtedly the most interesting and exciting one. Obviously, it poses an interesting theoretical challenge. We have already shown that the theoretical justification provided by the authors of the papers (the origin of the current/voltage drop is in the "symmetry breaking" by the external static magnetic field) does not stand a closer scrutiny (see Eq. 5.1). Here we suggest that the origin of the current/voltage drop resides only in the work function difference

---

[2]Measurements with the inversion of the magnetic field were also carried out in the Fu and Fu experiment, giving apparently consistent results. More on this later.

between the electrodes.

Let us start by recalling something about the objections related to the voltage drop (and electric field) that allegedly would build up instantaneously between the free ends of two materials with different work functions just after they are joined at one end. As already mentioned more than once, that voltage drop would prevent any thermionically emitted electron in the gap from spontaneously moving from a low work function to high work function free electrode surface. We have explicitly shown in Chapters 3 and 4 that no electric field, and thus no voltage drop, builds up between the surfaces at the free ends of two materials with different work functions as an instantaneous consequence of their having been joined at one end (and thus because of the formation of a depletion region along the junction).

We now analyze the Fu and Fu experiment in light of the theory outlined in those chapters. We have previously said that in their experiment, one will hardly have exactly $\phi_A = \phi_B$. Thus, one has to assume $\phi_A < \phi_B$ or $\phi_A > \phi_B$. Given that assumption and according to the analysis of Chapters 3 and 4, an equilibrium space charge eventually settles inside the bulb due to thermionic emission; under such an equilibrium, the plate with higher work function ends up with an excess of negative charge with respect to that with lower work function (which has an excess of positive charge). All this would happen even without the external static magnetic field. No built-in ("contact-generated") electric field between the free surfaces of the plates exists in advance that will prevent this final equilibrium. This means that, even without the external magnetic field, a voltage drop settles between the two terminals of the vacuum bulb and a current flows when the terminals are shorted through a load. When the permanent magnet is moved close to the bulb, the magnetic field can only modify the geometry of the space charge distribution. This is as though the physical geometry of the electrodes (relative distance, relative orientation) were actually modified. Note that, contrary to what may appear, in doing this action the static magnetic field $\mathbf{B}$ does not perform any work. We might also see the effect of the magnetic field in this case as an analog of the magnetoresistance phenomenon[3]. In the mathematical modeling carried out in the previous chapters, we mentioned how the geometry of the electrodes, their relative distance and orientation, influence the voltage drop, and consequently the amount of current flow when the electrodes are shorted, for instance, through the internal (high impedance) load of the measuring apparatus (electrometer).

This hypothesis could explain an interesting feature of the Fu and Fu experiment; namely, the switch from positive to negative (and vice-versa) of the values of current and voltage drop when the external magnetic field is reversed. Although in the Fu and Fu paper it is not explicitly stated, we guess that they zeroed the electrometer connecting its terminals to the bulb in the absence of any external magnetic field. However, this zero is not an actual zero. The reversal of the magnetic field does not actually reverse the flow of electrons from one electrode to the other; rather, it changes the space charge distribution increasing or decreasing the one-way current. However, the zero of the measurement apparatus being not the actual zero, such an increase or decrease above or below the relative zero is read as positive or negative by the electrometer.

5.2   More recent experiments by Fu and Fu

At the end of 2014 we received, as a personal communication, a report by Fu and Fu [140] on a further experiment conducted on the same set of vacuum bulbs employed for the tests described in the previous sections. In this latest series of experiments, however, Fu and Fu introduced two important changes in the experimental design with respect to the tests described in [136]. First, they excised the external static magnetic field. Secondly, they actually implemented and tested the basic scheme of a thermo-charged capacitor as the one sketched in Fig. 4.1; Chapter 4.

The way in which their original vacuum bulbs were manufac-

---

[3]The change of a material's resistivity with the application of a magnetic field.

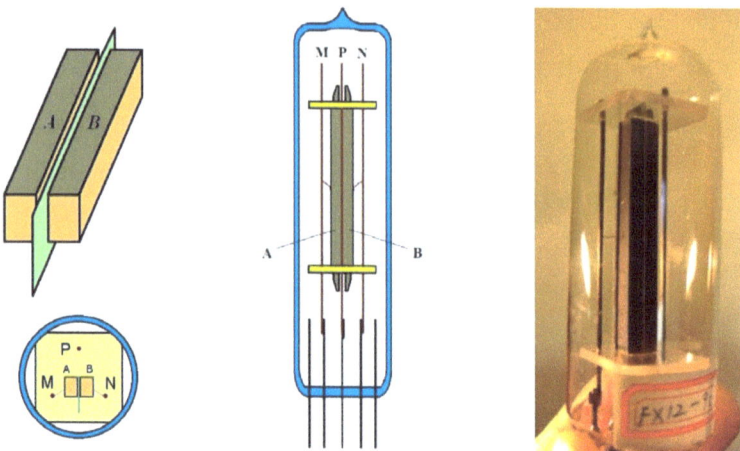

Figure 5.4: One of the vacuum tubes used by Fu and Fu [140]

tured made Fu and Fu able to use the Ag–O–Cs plates A and B of Fig. 5.1, connected in series, as the emitter and a cesium-coated molybdenum supporting rod facing the plates (P) as the collector of a simplified version of thermo-charged capacitor (see Fig. 5.4). A schematic of the experiment, taken from the Fu and Fu's report, is given in Fig. 5.5, where the vacuum tube, the shielding copper box and the measuring apparatus (electrometer Keithley 6514) are also visible.

Fu and Fu claim to have been able to measure a stable current exceeding 2×10^{-12} A and a voltage drop of the order of 100 mV. They also verified that both these values change sign once the electrometer's connections to the tube are switched over. Unfortunately, no dependence of the current/voltage output on the value of the external uniform temperature was investigated: the measurements described in the report have been taken at a single external temperature. Their results are extremely interesting and promising, and seem to give a first experimental corroboration of the theory behind the thermo-charging process reviewed in the present book. Obviously, we should be particularly careful, since what is at stake here is one of the most general and trusted laws of Nature, and further research (actually, much more research) is

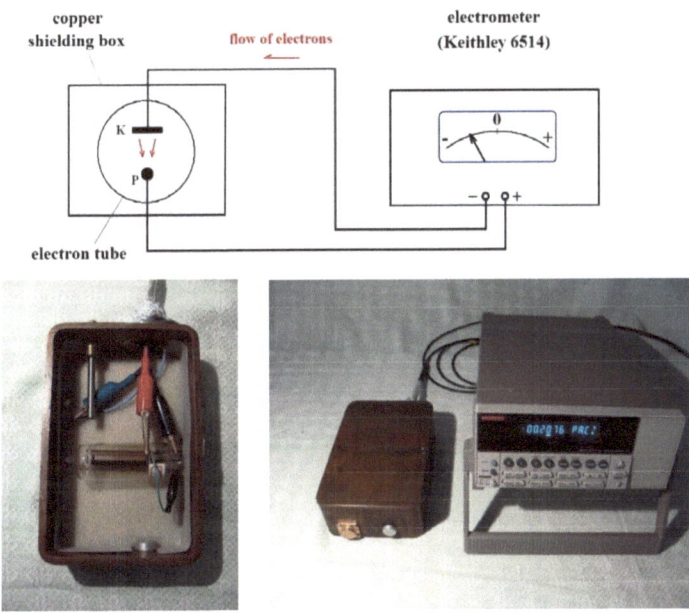

Figure 5.5: Schematic of the latest experiment carried out by Fu and Fu [140]

needed with stricter standards in terms of experimental design and protocol in order to be able to say the final word.

Chapter 6

Richardson's Nobel Lecture: thermionic emission and the second law of thermodynamics

This author remembers becoming aware for the first time of photo-electric and thermionic emission phenomena during his last years at secondary school (in the early 1990s). It did not take too much time to realize that these phenomena could have been at odds with the second law of thermodynamics[1]. Unfortunately, at the time he did not have the physical and mathematical background needed to investigate the matter further; the fact that he was not able to find any reference to this connection in all of the physics textbooks to which he had access, convinced him that, in fact, thermionic emission was not at odds with the second law. If he had been able to catch this connection, others had surely done the same long before him. If none of the physics books he found on the subject (even during the university years) describes a possible challenge to the

---

[1] Actually, it went like this: after having learned the basic principles of photo-electric effect, this author wondered whether some material could exist capable of emitting electrons when submersed in heat radiation at room temperature. He did not see any fundamental principle forbidding that process.

second law, it means that the photoelectric and thermionic emission phenomena trivially agree with it and the whole thing is not even worth mentioning.

This author dropped the issue and probably hid it in a remote corner of his brain. As a matter of fact, several years later, he had the chance to resume the issue and published few papers upon which most of the present book is based [126, 127, 128, 134, 135, 137].

During the years spent working on the theory behind the thermocharged capacitor, we have made an extensive bibliographic survey, almost failing to find any reference to a possible connection between thermionic emission and the second law, even among non peer-reviewed writings. It must be said that the near-complete absence in the scientific literature of any reference to this connection was and is still baffling.

The second law of thermodynamics is considered to be one the most fundamental and pervasive laws of Nature. It holds unconditionally, except for the microscopic world where small deviations are statistically common and have also been observed. As far as we know, the fact that macroscopic violations are considered impossible has not yet been rigorously and firmly proved (see discussion in Chapter 1): the second law inviolability stands mainly on the fact that no macroscopic violation has been observed thus far. For this reason, it is always important to check new physical phenomena against the second law in order to verify reciprocal consistency. In this context, it is somewhat puzzling that since the discovery of the emission phenomena there is a shortage of studies in the scientific literature on the connection between photoelectric/thermionic emission and the second law of thermodynamics. If there are physical processes that could pose a challenge to the second law, these are definitely the photoelectric/thermionic emission ones.

Nevertheless, not long ago (April 2015), we finally found an exception; a notable one indeed. The author happened to read the Nobel lecture "Thermionic phenomena and the laws which govern them" [11] by O.W. Richardson, the British physicist who discovered the law of thermionic emission that bears his name, and noticed that in the last pages Richardson explicitly mentions the second law of thermodynamics in connection with thermionic emission phenomena.

6.1 Richardson's Nobel lecture

Richardson's lecture [11] is a relatively short (13 pages, pp. 224–236), mildly technical and mainly historical review of his lifelong experimental and theoretical work that led to the formulation of the law that bears his name. It is obviously not meant to be a technical/scientific paper and should not be read as such, expecting strictness and exhaustiveness, but we found many interesting remarks and thought experiments, most of which amazingly predate what we have written in our papers (and in this book) on thermo-charged capacitors, with nearly the same physical formulation.

6.2 Black-body spectrum and photoelectric emission

For instance, on page 237, Richardson describes how he became aware of the fact that thermionic emission cannot be seen merely as an integrated photoelectric effect. He qualitatively describes how the more energetic tail of a black-body radiation (even at room temperature) is able to induce electron ejection from a metal, just like in the visible spectrum photoelectric effect. This is exactly what we have done quantitatively in our papers and in this book to support the possibility of harnessing energy from ambient heat through a sort of *thermo-photoelectric* effect:

> There is a very close relationship between thermionic and photoelectric phenomena. The photoelectric threshold frequency, the least frequency v_0 which will eject an electron from a given substance, is connected with the thermionic work function w_0 by the simple relation

$$w_0 = h v_0$$

where h is Planck's constant. This was established by experiments made by K.T. Compton and myself in 1912. *We know that any body in thermal equilibrium at any temperature T is surrounded by a bath of radiation in which the frequency distribution is given by Planck's formula. This formula puts no finite limit on the magnitude of the frequencies occurring; so that there will always be some frequencies present for which v_0 is greater than w_0/h. Such frequencies will eject electrons by photoelectric action; so that the temperature radiation alone will, by a kind of photoelectric effect integrated over the whole spectrum, give rise to an electronic emission which should increase with the temperature* [emphasis added]. In 1912 I showed that it followed from the principles of thermodynamics that this integrated photoelectric emission would follow Eq. (I) exactly with, possibly, a different value for the constant A. This conclusion was established by direct experiment later by W. Wilson, in 1917. Thermionic emission might thus well be an integrated photoelectric emission; only the absolute magnitude could decide. In 1912 there were no known data which would enable the magnitude of this integrated photoelectric effect to be ascertained, so, with the collaboration first of K.T. Compton and later of F.J. Rogers, I set about to determine the absolute values of the photoelectric yields of various substances as a function of frequency. With the help of these absolute values I was able in 1916 to calculate the electron emission from platinum at $2,000\,°K$ due to its complete black-body spectrum. The result showed that thermionic emission is at least 5,000 times, and almost certainly 100 million times, as large; so that thermionic emission cannot be merely an integrated photoelectric effect, although it has the same thermodynamic properties.

In this part of the lecture, Richardson does not pay any attention to the potential conflict of what he has just described with the second law: if it is possible to extract energy from the plain (UV spectrum) photoelectric effect, why should not be possible in principle to extract energy (although to a much lesser extent) from the photoelectric effect induced by the high-energy tail of the black-body spectrum at a single temperature on a device with two plates (anode and cathode) with different work functions? It could be objected that a UV spectrum photoelectric device works only when light impinges onto only one of the two electrodes, no matter whether they have different work functions, while in a "black-body spectrum" device both electrodes are inevitably affected by the same radiation (the device is submersed in the black-body radiation), no matter whether they have different work functions. A relatively simple way to settle the issue has been already presented in Section 4.2, and we shall not repeat its description here.

6.3   Thermionic emission and second law

The last three pages (pp. 234–236) of Richardson's lecture are where he eventually mentions the second law of thermodynamics. His main goal is not to raise the issue of a potential conflict between second law and thermionic emission. Rather, he uses the (universality of the) second law to show that the potential difference V_{12}, that builds up between two separate bodies (1 and 2) with different work functions (w_1 and w_2) due to the differential thermionic emission among them, should depend also on the distance between the two bodies. The reason is that small distances between charged bodies enhance the so-called *field extraction phenomenon*.

In order to prove his thesis, Richardson sets up the following thought experiment. His formulation is practically the same as that we made to describe the physical process at the basis of the thermo-charged capacitor functioning. In Richardson's own words:

> The existence of this field extraction phenomenon has a number of interesting consequences, one of which I will now mention. *If we consider an evacuated enclosure containing a number of bodies having different thermionic work functions*

w_1, w_2, *etc., they will not be in electrical equilibrium unless their surfaces are charged. The reason for this is that those with lower work functions would emit electrons at a more rapid rate than those with higher work functions. The condition for equilibrium to a first approximation, and one which covers the essential features of the phenomenon, is that there should be a certain field of electric force between the different bodies. This is such that, if the potential difference between any point just outside the body with suffix 1 and any point just outside the body with suffix 2 is* V_{12}, *then* $eV_{12} = w_1 - w_2$. V_{12} *is the contact potential difference between the bodies 1 and 2* [emphasis added]. There is nothing essential to the thermionic argument which depends on the shape, size or relative position of the bodies, and the result should be the same whether they are interconnected by other conductors or insulated from each other. The quantities such as V_{12} are thus intrinsic potential differences which are characteristic properties of the materials of which the conductors are made.

In the last two sentences, there is the claim he will disprove. First, he states that the potential difference V_{12} between any point just outside the body with suffix 1 and any point just outside the body with suffix 2 inside the evacuated chamber is equal to $eV_{12} = w_1 - w_2$ and V_{12} is the contact potential difference between the bodies 1 and 2 and then concludes: "*The quantities such as* V_{12} *are thus intrinsic potential differences which are characteristic properties of the materials of which the conductors are made.*" However, he then goes to show that the field extraction phenomenon complicates the situation:

The field extraction phenomenon requires a modification of this conclusion. To simplify the argument I consider only two bodies, those with suffixes 1 and 2. Some portion of each of them is bounded by a plane surface, and the bodies are arranged so that these plane surfaces are parallel to one another and a distance x apart. The more distant parts

of the bodies may be united by an electric circuit which
includes a galvanometer. *When x is considerable, there is
equilibrium and no current passes through the galvanometer,
because the excess electrons emitted by the more electropositive body* [lower work function] *are kept back by the potential
difference V_{12} and this equilibrium is practically unaffected by
the small force eV_{12}/x* [emphasis added]. But now suppose
x to become very small, let us say comparable with atomic
dimensions. The force eV_{12}/x now becomes large and will be-
gin to extract electrons from the more electronegative body
[higher work function]. *This upsets the equilibrium, which
is restored by a current passing through the galvanometer.
But this is a perpetuum mobile: the current can be made
to do useful work. It consumes nothing and the apparatus
has no moving parts. If it is argued that it may be tapping
some source of heat, at least it must be a perpetuum mobile
of the second kind, since it works at a constant temperature*
[emphasis added]. What is the answer to this riddle? I say
it is this: the contact potential difference V_{12} is not completely independent of the distance between the two bodies.
*When this distance becomes small, V_{12} diminishes, and this
diminution takes place in such a way that the additional electron current from the more electropositive body which reaches
the more electronegative body owing to the reduced value of
V_{12} is just equal to the electron current which is extracted
from the more electronegative body by the field. In particular
when the bodies are in contact, V_{12} falls to zero or at any
rate to a quantity of the order of the thermoelectric magni-
tudes* [emphasis added]. Well, this seems to correspond to
the actual properties of the contact difference of potential,
and I think it clears up an old difficulty in connection with
it.

Here, Richardson plays the second law against the idea that
the thermionically generated potential difference V_{12} does not depend upon the distance between bodies 1 and 2. He does so in a
somewhat bizarre way, though. He seems to consider the equilib-

rium that gave rise to V_{12} like something externally imposed on the system with the inescapable urgency to be restored (e.g. by means of the current passing through the galvanometer), whatever happens. As a matter of fact, V_{12} is the result of the *effective* electron flux between the body with lower work function (electropositive) and the body with higher work function (electronegative), and it is quite clear that if there is an opposing force (field extraction) driving the electrons the other way around, a sort of new equilibrium is dynamically reached, where the potential difference (ultimately generated by the new *effective* electron flux) is lower than before. The entire process with these two opposing "forces" is completely local; it only happens across the close surfaces of the two bodies and there is no need to involve currents passing through an external circuit connecting the most distant parts of the bodies. The appeal to the second law seems to be like using a machine gun to kill a mosquito. We believe that the disturbance of field extraction to V_{12} becomes non-negligible only when the distance between the surfaces is, as Richardson notes, comparable to atomic dimensions.

What Richardson calls "the answer to the riddle", namely that V_{12} goes to zero when x goes to zero, is an interesting result that is apparently overlooked in the literature on contact potential between metals. We have already noted that in almost all the textbooks on contact potential, it is written that when two uncharged metals with different work functions ($w_1 > w_2$) are joined together, a macroscopic potential difference (equal to $V_{12} = (w_1 - w_2)/e$) instantaneously builds up between any point on the surface of metal with suffix 1 and any point on the surface of metal with suffix 2 far from the junction, like if charges spread over the surface of the two metals (negative charges over the more electronegative metal and positive charges over the more electropositive metal). We have shown in the previous chapters that this does not stand logical scrutiny: there is no macroscopic potential difference between any two points on the whole surfaces the two metals when they are physically joined along a small portion of their surface and the two metals remain overall uncharged. Charge displacement upon contact remains confined in the microscopic layer between the contact surface and across that layer, and only across that layer, we have that $V_{12} \neq 0$.

There is another baffling statement made by Richardson in the last part of his lecture: giving the premises of his thought experiment with two separated bodies with more distant parts united by an electric circuit which includes a galvanometer, he says "*When x is considerable, there is equilibrium and no current passes through the galvanometer, because the excess electrons emitted by the more electropositive body are kept back by the potential difference V_{12} and this equilibrium is practically unaffected by the small force eV_{12}/x*". However, the connection between the two bodies through an electric circuit that includes a galvanometer is another way to put them in contact; if the conclusion of the thought experiment is right, i.e. $V_{12} = 0$ when $x = 0$, then V_{12} should be equal to zero just from the outset, well before the experimenter ideally brings the free plane surfaces of the two (already connected) bodies close to one another and eventually in contact ($x = 0$). Instead, Richardson states that "*[...] there is equilibrium and no current passes through the galvanometer, because the excess electrons emitted by the more electropositive body are kept back by the potential difference V_{12} [...]*". Furthermore, if it is true that V_{12} falls to zero just by connecting the two bodies with the electric circuit (including the galvanometer), then there is no longer a potential difference that keeps back the "*the excess electrons emitted by the more electropositive body*" between the two planes surfaces of the bodies and a steady current originating from thermionic emission (and thus, from ambient heat) should flow through the galvanometer ($V_{12} \simeq 0$, $R_{12} \simeq 0$ and $I_{12} \neq 0$, where R_{12} is the ohmic resistance of the circuit). This is a violation of the second law as well. It would be interesting to know why he did not make any mention to this. As we have already said, Nobel lectures are not meant to be technical papers and one should not expect exhaustiveness: this second "riddle" was with very high probability clear to Richardson but, for some reasons known to him, he felt that it was superfluous to dwell on this.

Chapter 7

On the prediction of asymmetric ageing in Einstein's 1905 paper

[...] yet his tragedy lay precisely in an almost morbid lack of self-confidence.
He suffered incessantly from the fact that his critical faculties transcended his
constructive capacities. In a manner of speaking, his critical sense robbed him
of his love for the offspring of his own mind even before they were born.

"Paul Ehrenfest in Memoriam",
Albert Einstein, *Out of my later years* (1934) [16].

Among the most bizarre conclusions of Einstein's special relativity is what is now widely known as the "twin paradox": it is a thought experiment that, in its more popular version, involves two identical twins, one of whom makes a journey into space in a spaceship traveling at a velocity close to that of light c, and returns home to find that the twin who remained on Earth had aged more. This result appears puzzling (hence the name) because each twin sees the other twin as moving, and so, according to a hasty application of the principle of relativity and time dilation, each should paradoxically find the other to have aged more slowly. The apparent paradox is usually resolved appealing to the lack of symmetry between the frames of the two twins: the traveling twin's trajectory

involves two different inertial frames, one for the outbound journey and one for the inbound journey, while the inertial frame of
the Earth-bound twin stays the same[1].

The seemingly strange (peculiar) behavior of two previously
close and synchronized clocks, one of which makes a round trip
at high speed, has been pointed out by Einstein already in his
1905 relativity paper. Let us now fully quote his brief and elegant
description of the phenomenon [3]:

> Let us now consider that a clock which is lying at rest in the
> stationary system gives the time t, and lying at rest relative
> to the moving system is capable of giving the time τ; suppose
> it to be placed at the origin of the moving system k, and to
> be so arranged that it gives the time τ. How much does the
> clock gain, when viewed from the stationary system K? We
> have,
>
> $$\tau = \frac{1}{\sqrt{1 - \frac{v^2}{c^2}}} \left(t - \frac{vx}{c^2} \right) \quad \text{and} \quad x = vt,$$
>
> $$\therefore \tau - t = \left[1 - \sqrt{1 - \frac{v^2}{c^2}} \right] t.$$

Therefore the clock loses by an amount $\frac{1}{2}\frac{v^2}{c^2}$ per second of
motion, to the second order of approximation.

From this, the following peculiar consequence follows. Suppose at two points A and B of the stationary system two
clocks are given which are synchronous in the sense explained
in § 3 when viewed from the stationary system. Suppose the
clock at A to be set in motion in the line joining it with B,
then after the arrival of the clock at B, they will no longer
be found synchronous, but the clock which was set in motion
from A will lag behind the clock which had been all along at
B by an amount $\frac{1}{2}t\frac{v^2}{c^2}$, where t is the time required for the
journey.

---

[1]This lack of symmetry is particularly evident if one represents the frames
geometrically in the Minkowski space-time diagram.

*We see forthwith that the result holds also when the clock
moves from A to B by a polygonal line, and also when A and
B coincide. If we assume that the result obtained for a polygonal line holds also for a curved line, we obtain the following
law. If at A, there be two synchronous clocks, and if we set
in motion one of them with a constant velocity along a closed
curve till it comes back to A, the journey being completed in
t-seconds, then after arrival, the last mentioned clock will be
behind the stationary one by $\frac{1}{2}t\frac{v^2}{c^2}$ seconds.* [emphasis added]

This original conclusion and analogous derivations by Einstein
and other authors in the subsequent years have generated a plethora
of analysis, criticism and reply to criticism. It is neither our intention nor in the range of our ability to survey them here. For
the interested reader, it is enough to make a simple bibliographic
search of the history of the twin paradox: he/she will find hundreds
of entries about the topic.

In this chapter we humbly join the community of scholars[2] who
think that the logical structure of Einstein's theory of special relativity is not exactly what one would define as unproblematic. The
aim of this chapter is to provide evidence, with two easily understandable thought experiments, that if both postulates of the theory of special relativity are assumed to hold true concurrently, then
the prediction of asymmetric ageing made by Einstein in his 1905
relativity paper is in fact not compatible with them; and the fact
that time dilation (which is intimately connected to "asymmetric
ageing") seems to have been experimentally confirmed provides,
paradoxically, a refutation rather than a confirmation of the theory of special relativity, at least as interpreted today. According
to our analysis of the logic behind special relativity, no time dilation should occur under any circumstances at all. Hence, this
points to a twofold contradiction: time dilation seems actually to
have been observed in several experiments (and should not; first
contradiction), but only in some and not in all the circumstances
allegedly predicted by the theory (if the theory were actually cor-

---

[2]The most influential of whom has been without doubt Professor Herbert
Dingle.

rect, time dilation should occur in all the circumstances predicted by the theory itself; second contradiction).

In the following section, we introduce the first of two thought experiments: it involves two satellites orbiting about a planet. We show that it is possible to set up a perfectly symmetrical version of the twin paradox. We also critically evaluate the possibility of any resolution within the theory of special and general relativity on the basis of fundamental physics principles, like the equivalence principle, and also by considering GPS satellite data.

In Section 3 we introduced the second thought experiment. It concerns the measurement of the velocity of two light pulses emitted from two reference frames relatively moving with respect to a third frame (and with respect to one another). We show that if both postulates of special relativity are assumed to be true concurrently, the Lorentz transformations cannot describe real, physical changes of space and time and, strange as it may appear, the term "velocity", when ascribed to light, cannot have the same meaning as that we intend in classical mechanics for ordinary material bodies: in a precise sense, light behaves differently than an "ordinary" physical object. A possible explanation of why it is so (still a mere speculation, though) is also given.

In the last section we briefly provide the reader with references to and quotes from some papers by Mendel Sachs, one of the relatively few distinguished scholars who championed the position that there are no material consequences, like asymmetric ageing, implied by the space-time transformations of relativity theory.

7.1   Twin Satellites Paradox

Consider two identical satellites orbiting on two circular paths with nearly the same radius about a non-rotating planet (Fig 7.1). The orbits are close to one another; one satellite moves clockwise while the other moves anti-clockwise. Therefore, their orbital velocities are equal in magnitude ($v \lesssim c$) and opposite in direction.

Now, consider their respective reference frames, S_1 and S_2. It

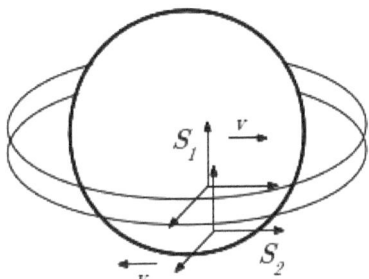

Figure 7.1: The twin satellites scenario: satellites S_1 and S_2 move around a planet in nearly equal circular orbits, with opposite velocities (same speed $v \lesssim c$).

is not difficult to realize that *both* S_1 and S_2 are inertial frames (free falling systems): in both frames Newton's laws of dynamics hold true, at least if we take into account displacements from the origin of S_1 and S_2 small compared to the orbit radius (in order to have negligible tidal forces). An observer in S_1 (S_2) may well believe he/she is at rest and see S_2 (S_1) passing by at uniform velocity (v). This scenario is *perfectly symmetrical*; hence if we follow Einstein's argument quoted above in italics, in what frame does a clock lag behind the other?

7.1.1 Discussion

It is clear that for the resolution of this puzzle, we can no longer rely on the lack of symmetry between the space-time paths of the two systems: they are fully and trivially equivalent.

It would have been interesting to know what would have been Einstein's reply to this paradox. As far as we know, there is no mention of this thought experiment in his scientific oeuvre, even after the "happiest thought of his life", namely the discovery of the equivalence principle, and even after the mathematical machinery of general relativity was fully developed. Einstein himself devoted some effort in 1918 [9] to replying to objections against the classical twin (or clock) paradox. He addressed the paradox within general relativity and ultimately ascribed the non-symmetrical outcome to the acceleration suffered by only one of the two twins (clocks) before reunion. We now know that acceleration does not play any role in the resolution of the twin paradox: many theoretical arguments [117] and experimental data (e.g., muon decay in storage rings[3]) show that clearly.

At first sight, it appears that with the twin satellites scenario, we cannot resort to general relativity either. The presence of the planet with its gravitational field is only a ploy to bend S_1 and S_2 paths: the whole system (S_1, S_2 and the planet) is still symmetric. Even if some alleged resolution is provided according to which the ageing comes out to be asymmetric, what is the criterion to make a distinction between two completely identical situations? Moreover, observers in S_1 and S_2 (suppose them unable to see the central planet because it is too small and dim) physically perceive themselves as being in inertial frames; according to the equivalence principle, there is no internal physics experiment they can perform to detect that they are actually falling (orbiting) in a gravitational field. They are fully entitled to describe what they see with the same laws of physics that hold in inertial frames infinitely far from planets and stars and, in particular, with the machinery of the theory of special relativity.

To our knowledge, the twin satellites paradox was suggested for the first time in 1972 in a paper by Holstein and Swift [29]. In the introduction of that paper, the authors describe the thought experiment and then dismiss the relevance of the paradox that comes out of it as follows:

---

[3]Special relativity is applied with no hesitation or afterthought to explain successfully time dilation in muon decay inside storage rings, where the timekeepers—the muons—travel at high speed around a ring and undergo high centripetal accelerations. Almost nobody seems to worry about the fact that, if the equivalence principle is true, then the high centripetal acceleration suffered by the muons should, as required by general relativity, sensibly contribute to time dilation. However, there is no trace of this contribution in the measured time dilation [141].

Two satellites are in counter revolving circular orbits about the Earth. At $t = 0$ they meet over the North Pole and half a period later they meet again over the South Pole. If the Earth were shrunk to a point, an astronaut in each satellite could observe the other during the time between the meetings. Each astronaut would be in free fall; and, if he observed only the other satellite, he could claim that he was at rest in an inertial frame while his counterpart in the other satellite traveled out and returned. Since he is at rest, he could apply his limited knowledge of relativity and conclude that the other astronaut, traveling as he is, will be the younger one when they meet again. Yet the observations of the two astronaut are entirely symmetric. *An observer on the Earth would conclude that the two astronauts age identical amounts and, by symmetry, this must be the correct answer.* [emphasis added]

We do not think that the short observation emphasized at the end of the quotation is sufficient to dismiss definitively the paradox and clear all doubts about special relativity. First, if we accept that conclusion, we trust the only non-inertial observer on the stage: sometimes relativists say that special relativity cannot be applied in non-inertial reference frames; sometimes they say it can. In our opinion, this way of doing seems to depend too much upon the circumstances.

And even if with "*An observer on the Earth*" the authors meant the abstract Earth-centered inertial frame[4], this frame is likewise a free-fall reference frame; in this case, orbiting about the Sun. Why can we therefore apply special relativity in one free falling frame (that about the Sun) but not in the other (that about the Earth or a planet)? Secondly, if we take as valid the logic used by the authors in the quoted (and emphasized) paragraph, then we fail to see why it cannot be applied also to the following scenario too: we have two relatively moving inertial frames K_1 and K_2

---

[4]The Earth-centered inertial coordinate frame (or ECI) has its origin at the center of mass of the Earth, but does not rotate with it (i.e it is fixed with respect to the stars). The time in ECI is adopted as the basis for GPS time.

and a third inertial frame K_3 taken as the standard of rest for the sake of argument. Suppose further that K_1 moves eastward with constant speed v and K_2 westward with the same speed v, both with respect to K_3 and both in linear motion. In that case *an observer on K_3 would conclude that K_1 and K_2 age identical amounts and, by symmetry, this must be the correct answer.* If this is correct, then the fact that according to special relativity K_1 sees K_2 as ageing slower (and vice-versa) is pure illusion. According to the very same principle of relativity, even the conclusion of K_3 that both K_1 and K_2 actually age (no matter if identical amounts or not) is pure illusion. This is exactly what we are stressing here.

In the remainder of the paper, the authors go on proving with the tools of general relativity that a satellite in a circular orbit and a satellite in an elliptical, parabolic or hyperbolic orbit age differently, even though neither satellite experiences acceleration. In particular they say:

> *The standard special relativity arguments, while leading to the correct conclusion when the traveling twin returns by firing retrorockets, fails* (sic) *when both observers are in free fall in a gravitational field.* [emphasis added] There is something peculiar about the fact that free fall in a gravitational field can turn an inertial frame around. To put it in another way, in a gravitational field, two inertial frames—frames in which Newton's laws are valid—need not to be in uniform relative motion. In order to analyze motion in gravitational field we must use general relativity.

They suggest that special relativity does not apply to cases where both observers (satellites) are in free fall in a gravitational field. However, the theory of special relativity was developed for *inertial reference frames* and in a Universe filled with mass, is there any inertial reference frame that is not also a "free fall" reference frame? If special relativity is physically valid or if it claims to explain what actually happens in the world around us, it must be valid for actual inertial reference frames, all of which are also "free fall" frames, and not only for the abstract concept of inertial

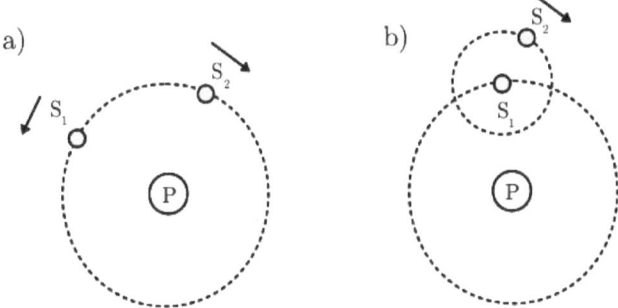

Figure 7.2: Scenario a): S_1 is *not allowed* to apply special relativity to determine time dilation in S_2 because this generates a paradox (twin satellites paradox) and because, according to many scholars, when both S_1 and S_2 are observers in free fall in a gravitational field standard special relativity fails. Scenario b): S_1 is *allowed* to apply special relativity to determine time dilation in S_2. As a matter of fact, this is exactly what happens with the determination of the velocity time dilation in GPS satellites (P is the Sun, S_1 is the Earth and S_2 is the GPS satellite). We fail to understand why the arguments that dismiss the applicability of special relativity to case a) do not hold also for case b). After all, both S_1 and S_2 in b) are also observers in free fall in a gravitational field.

reference frame (which, by the way, does not physically exist). Furthermore, if standard special relativity fails when both observers are in free fall in a gravitational field, as the authors maintain, then it must also fail when an observer is on the Earth and the other is on an orbiting satellite, since both the Earth and the satellite are in free fall in the gravitational field of the Sun, and the Sun in free fall in the gravitational field of the galaxy, and so on. However, this seems not to worry those people too much who apply standard special relativity to GPS satellites from the Earth to determine, for instance, the velocity contribution to time dilation and also consider what they measure as an amazing experimental confirmation of the theory (see Fig. 7.2). If one accepts that special relativity can be successfully applied from the Earth to explain the velocity time dilation on a GPS satellite but cannot be applied from one satellite to the other to avoid the paradox, then the very same person cannot fail to notice that the paradox inexorably resurfaces if one considers the Earth and, for instance, Mars as the new twin satellites, this time both orbiting about the Sun.

In fact, an experimental realization of the twin satellites scenario comes from the Global Positioning System (GPS) [62][5]. This system consists of a network of 24 satellites in roughly 12-hour orbits, each carrying atomic clocks on board. The orbital radius of the satellites is about four Earth radii. The orbits are nearly circular. Orbital inclinations to the Earth's equator are about 55°. The satellites have orbital speeds of about 3.9 km/s in a frame centered on the Earth and not rotating with respect to the stars. Every satellite has on board four atomic clocks marking time with an error of a few ns/day. From every point of the Earth's surface at least four satellites are visible at any time. The theory of general relativity predicts that clocks in a stronger gravitational field will tick at a slower rate. Thus, the atomic clocks on board the satellites at GPS orbital altitudes will tick faster by about 45900 ns/day because they are in a weaker gravitational field than atomic clocks on the Earth's surface. The velocity effect predicts that atomic clocks moving at GPS orbital speeds will tick slower by about 7200 ns/day than stationary ground clocks. Therefore, the global prediction is a gain of about 38700 ns/day. Rather than having clocks with such large rate differences, *the satellite clocks rates were reset before launch (slowing them down by 38700 ns/day) to compensate for these predicted effects.* The very rich data show that the on-board atomic clock rates do indeed agree with ground clock rates to the predicted extent. Obviously, daily corrections from the ground segment managing the GPS system are needed in order to compensate for minor rate changes, probably due to small orbital perturbations, random noise, etc. Note that the pre-launch rate compensation is independent of frame or observer considerations. In particular, pre-launch rate compensation is the same for every satellite, and *no relative time dilation* seems to apply to any couple of GPS satellites. Namely, no couple of GPS satellites seems to go out-of-sync because of their relative velocity time di-

---

[5]Some portions of the following text on the Global Positioning System, the most technical ones, are quoted verbatim from the original reference.

lation. The reason is surely because there actually is no relative time dilation at all: it would be unthinkable that the satellites actually went out-of-sync, but that this loss of synchronization is not detectable in the reference frame of the Earth.

All this seems to suggest that time dilation is not a real, physical effect affecting a moving observer in its own reference frame, but an effect that shows up when observers in different reference frames gather information through electromagnetic (em) signals. In some sense, the effect is born "in the middle". Ground receivers measure the time coming from satellites through radio signals and thus, as far as we know, one cannot be completely sure whether the change in rate is actual and physical on board the satellites in orbit or it comes as a not-yet explained side effect in the transfer of em waves between the satellite and the ground station (see the speculation discussed at the end of Section 7.2).

It would be interesting to bring back to the ground an intact and still-working GPS satellite and compare the overall clock cycles recorded by the on-board atomic clock[6] to those measured on the Earth through the radio signals sent by the clock during all the time it was in orbit. We think that this would be the real crucial test.

7.2   On the velocity of light, time dilation and length contraction

Until now, the ascription to the physical phenomenon of light of mechanical properties like velocity in its commonplace meaning (the same meaning, for instance, we intend when we describe the motion of a macroscopic material object in space) has been unproblematic. It has not even been a reason of inquiry. Yet, are we sure that this is as unproblematic as it appears? We shall show that if we accept the two postulates of the theory of special relativity as concurrently true, then we must agree that the meaning

---

[6]If ever there were something on-board to keep track of the cumulative clock cycles.

of the words "velocity of light" is not as obvious as it is believed to be; and, further, Lorentz transformations cannot describe real, physical changes of space and time, unless we are ready to accept that the *law of the excluded middle* is physically false.

In what follows, our focus is exclusively on kinematics and we do not consider the alleged dynamical effects of relativistic velocities. In fact, what we discuss is also true for non-relativistic velocities ($v \ll c$) of the reference frames.

Suppose that, for the sake of argument, when we talk about the velocity of light the meaning of the word "velocity" is exactly the same we intend when we talk about the velocity of an ordinary material body, like a ball, moving in space. Namely, *a pulse of light is like a physical object that maintains its own identity when seen from different inertial reference frames and its velocity is the space traveled by that physical body divided by the time taken to cover this space* (Assumption 1, or A1). Moreover, according to the second postulate (P2) of special relativity, as measured in any inertial frame of reference, light is always propagated in empty space with a definite velocity c that is independent of the state of motion of the emitting body. In what follows we also make implicit use of the first postulate of special relativity (P1).

Consider now the following thought experiment. We are in a reference system S_2 moving at uniform velocity v (as said before, even with $v \ll c$) towards a reference system S_1 assumed to be the standard of rest. In S_1 a pulse of light is generated in the direction of the incoming S_2. According to P2, the velocity of light must be the same (c) in S_1 and S_2. Lorentz transformations come to our help and, seen from S_1, the units of length and time in S_2 change as much as needed in order for S_2 to measure a velocity equal to c (see, for instance, [8]). However, if we now let A1 come in, we see that the unit changes given by the Lorentz transformations must be real, physical changes in S_2. If light is like an ordinary material body maintaining its own identity in different reference frames (e.g. a rubber ball), in order for S_2 actually to measure on the same object the very same velocity as in S_1, it is necessary that the change in the units of length and time be real and physical in S_2, not only apparent from S_1. Also note that also the relativistic explanation of magnetic forces (Purcell's basic explanation of magnetic forces, see Born [22]) demands Lorentz contraction to be a real contraction, see Appendix A.

However, a serious contradiction now appears. Suppose that a second source of light S_3 comes into play that moves in the same direction of S_2 with a velocity $w \neq v$. According to P1, S_3 can be now considered the standard of rest and the same argument as before can be applied to S_2. Given A1, P1 and P2, in order for S_2 to measure a velocity equal to c for both light pulses emitted by S_1 and S_3 it is necessary that two real, physical, but mutually incompatible unit changes (two distinct Lorentz factors $\gamma = 1/\sqrt{1 - v^2/c^2}$) take place at the same time in S_2. This is logically and physically implausible (i.e the law of the excluded middle).

It must be noted that if we consider the same thought experiment in the framework of a relativity theory, like that of Lorentz, where the first postulate does not hold and a preferred frame exists where light propagates, this contradiction disappears. In that case, there is only one Lorentz factor: that related to the velocity of the reference frame S_2 relative to the preferred frame.

After some thought, it should be clear that the import of the reasoning above is twofold: if we accept as concurrently true the two postulates of special relativity and assumption A1, then 1) Lorentz transformations do not describe actual, physical changes of space and time; if, instead, we accept as concurrently true the two postulates of special relativity but not assumption A1[7] then 2) a pulse of light does not maintain its own identity from a reference frame to another reference frame and Lorentz transformations are superfluous. There is no longer any need to talk about space and time changes in order to have the same beam of light with the same velocity in different reference frames: a beam of light seen from two different reference frames is no longer the same object.

---

[7]The third case, A1 true and special relativity postulates not physically true, is not explicitly dealt with here. We guess that it directly points to a Lorentz type of relativity theory, where the first postulate does not hold and a preferred frame exists where light propagates at speed c.

More on this later. If conclusion 1) is true, then asymmetric ageing cannot occur: if for every infinitesimal interval of time dt Lorentz transformations do not describe a real, physical change in time rate; then no real, physical asymmetric ageing can occur after the integration of dt over a round trip. Furthermore, if conclusion 2) is true, then Lorentz transformations do not respond to any need of explanation of the actual physical phenomena, and thus there is no reason why the mathematical "consequences" they bring (changes of space and time) must be physically true.

As a matter of fact, could the concepts of "real, physical length contraction" or "real, physical time dilation" have any meaning at all if not intended with respect to a *preferred* space and a *preferred* time? Moreover, within special relativity, what does "unit length contraction" really mean, or, which is the same[8], "space contraction" in moving reference frames if "space" does not actually belong to any reference frame in particular, the reference frames being abstract human conventions? Moreover, how can the choice of a reference frame, which is a human construction from which we observe the physical phenomena, have objective, physical consequences on bodies in other reference frames, as seems to happen in special relativity? We invite the reader to ponder these issues further.

Let us now linger on the second conclusion. As a matter of fact, it is not that strange. When one measures the velocity of a macroscopic body, like a ball, he/she physically interacts with the object; such an interaction is a mild one: the experimenter uses light to see where the ball is at a particular instant of time. However, if one wants to measure the velocity of light, things are different. If we want to measure in our reference system S_2 the velocity of a beam of light emitted in S_1, our experimental apparatus must considerably interact with that beam of light: for instance, in the Foucault method, a beam of light has to be reflected multiple times between the two mirrors of the measuring apparatus. Thus, what we are

---

[8]The reader should recall, for instance, that in Purcell's basic explanation of magnetic forces, what really contracts is the space between the charges seen as moving.

measuring may not be the velocity of the original beam of light, but rather the velocity of the light that has interacted with the matter in our reference frame (the experimental apparatus is co-moving with S_2). In some sense, what we are measuring is the velocity of the light *generated in our reference frame* after the interaction of the original beam of light with the matter in *our* reference frame (the measurement apparatus) and, according to Maxwell's equations, its velocity in *our* reference frame is always c. Can we really believe that a beam of light maintains its own identity (whatever that means) even after a simple mirror reflection such as a ball rebounding from a wall does? Even after a single mirror reflection, what we see is the light generated by the matter of the mirror after the interaction with the original beam; this happens according to the Maxwell's equations in the reference frame of the mirror.

Here we want to be even more extreme and say that the very same source of light or of em waves actually generates different beams at the same time when that source is observed from different reference frames: there is no unique beam (object) observed from two different inertial reference frames as it may happen with a ball moving in space. If we observe a macroscopic physical object from two relatively moving reference frames we see the *same* object as having different velocities; but if we observe a source of light from two different reference frames, we do not see the *same light beam* (object), but two different physical phenomena. If we accept that Maxwell's equations hold in every inertial reference frame and describe properly the generation and propagation of em waves[9], then this conclusion seems inevitable. Consider a

---

[9]As a matter of fact, the possibility that Maxwell's equations do not provide a fundamental and complete theory of light generation and propagation and that they are only phenomenological laws is obviously not so remote: remember that light is a quantum phenomenon. What turns out to be most puzzling to this author, beside Einstein's way of thinking expressed in the epigraph to this chapter, is the fact that he published in the very same year two papers, on the photoelectric effect and on special relativity, which are, in some sense, actually at odds with one another. With the theory of special relativity, Einstein apparently succeeded in making Maxwell's equations and the principle of relativity compatible. However, before submitting that paper, he submitted (and published) the paper on his interpretation of the photoelectric

source of em waves O, like that depicted in Fig. 7.3: for the sake of simplicity, O is a wire where electrons move according to the velocity law $v(t) = v_0 \sin(\omega t)$. The source O is at rest in the reference frame S. The application of Maxwell's equations to this physical setup, with the boundary conditions given by, among other things, the electrons' velocity $v(t)$, should allow us to obtain the equations of the em waves generated by O. In the reference frame S, the speed of these em waves is obviously equal to c, according to the Maxwell's equations. Consider now the same source O seen from a different inertial reference frame S' moving at a relative velocity w with respect to S. According to an observer in S', the kinematics of the electrons in the wire O is now different. It is not important to try to figure out what it is. If the observer in S' applies Maxwell's equations with these new boundary conditions, he/she obviously obtains em waves moving at speed c in S'. Here we contend that these two kinds of em waves are not trivially the same em wave seen from two different reference frames, but rather two distinct physical phenomena, despite the fact of their being generated by the same source O.

This apparently strange conclusion can be corroborated by the following simple thought experiment. Consider a permanent magnet at rest in a reference frame S and generating a static but non-

---

effect. This interpretation represents a crucial step toward quantum mechanics and, above all, suggests/supports a physical model of light that is incompatible with Maxwell's equations. With the paper on the photoelectric effect he actually gave up Maxwell's theory of light, but with the paper on relativity he redeemed Maxwell's equations from the inconsistency with the principle of relativity. Every other physicist would have not slept quiet. Abiko [109] maintains that special relativity was constructed, or at least thought of, as a theory applicable beyond the realm of the applicability (and the validity) of Maxwell's theory. This is the reason why, according to Abiko, Einstein felt the need to introduce a separate postulate on the constancy of light, a constancy that would have been otherwise derived by assuming the validity of Maxwell's equations and the first postulate. We do not know whether this actually corresponds to the historical facts, but it is surely suggestive. One thing, though, is hardly deniable: what seems to have been, for Einstein, the main drive to the derivation of special relativity is the need to reconcile classical electrodynamics (governed by Maxwell's equations plus Lorentz force) and the principle of relativity. It is exactly with reference to this problem that Einstein's 1905 paper begins.

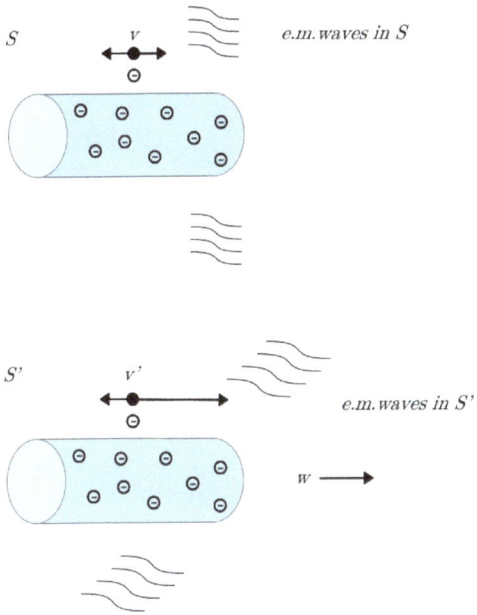

Figure 7.3: Piece of wire with an oscillating current, seen from two inertial frames. Frame S is the frame where the wire is at rest, while S' is the frame where the wire is moving with velocity w. In both inertial frames, we can apply Maxwell's equations but with different boundary conditions (the electrons have a different law of motion). In both frames, an electromagnetic radiation originates that propagates at the speed of light c. However, these two radiations are not simply the same ("same object") seen from two different "points of view", but in fact two distinct physical phenomena.

spatially uniform magnetic field (Fig. 7.4). If we now consider a different *abstract* reference frame S', moving with constant velocity w with respect to S, then at any point in S' there is a magnetic field changing with time[10]. This means, according to Maxwell's equations applied in S', the presence of electromagnetic waves spreading over the space. However, S' being an abstract reference frame of which S may well be unaware, any observer in S does not detect any em waves around. As a matter of fact, there is an infinite num-

---

[10]Depending upon the spatial distribution of magnetic field $\mathbf{B}$, we not only have $\frac{\partial \mathbf{B}}{\partial t} \neq 0$, but even $\frac{\partial^2 \mathbf{B}}{\partial t^2} \neq 0$.

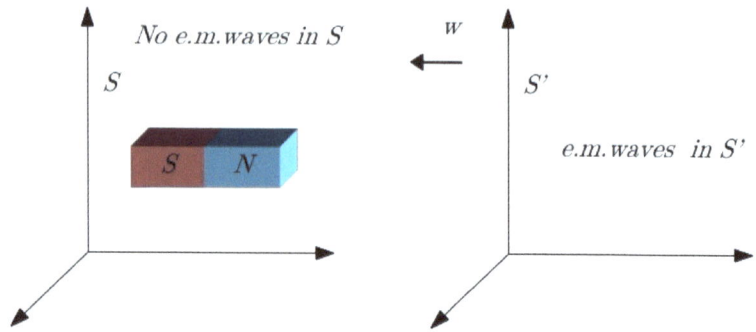

Figure 7.4: Permanent magnet thought experiment.

ber of abstract systems like S' of which S is not aware. According to the abstract reference frame S', there should be an energy dissipation due to the em waves propagating in space. Where does this energy come from?

The situation is as follows: an observer in S does not detect any em waves around, while an observer in S' does. This holds for an infinity of observers like S'. If an observer in S''' switches from S''' to S', e.g. by changing his/ her velocity with respect to S, then he/she instantaneously stops detecting the em waves typical of S''' and starts detecting the em waves typical of S' (if the observer switches from S''' to S, then he/she stops detecting any em waves whatsoever).

What more clear example than this to show that, if we accept Maxwell's equations as good, then the light generated by the same single source is not the same physical thing when observed from two different reference frames ? As a further example, if one considers a single charge at rest in a uniformly accelerated frame K, then an observer at rest in K will see no em waves spreading in space, while an observer in an inertial frame will see the charge as generating an electromagnetic radiation. Incidentally, this is exactly the way in which some authors have addressed within general relativity the equivalence principle paradox, namely the paradox of a radiating/non-radiating charge in a gravitational field (see [24, 34, 118]).

The theory of special relativity and Lorentz transformations would thus be mathematical expedients devised to force a beam of light, intended as a single material object maintaining its own identity, to have the same velocity c when seen from two different reference frames; the consequences they bring (length contraction, time dilation, mass increase, to name a few) might be as physical and real as the effects of a distorting lens on images: if we see two electric charges through a lens appear closer than they are, they surely do not exert for this reason a greater repulsive force one another. The analogy with the laws of optics is not to be taken too literally. The laws of optics tell us how light actually behaves when passing, for instance, through a lens; while, according to our interpretation, Lorentz transformations do not describe what we will see if we were able to observe by sight objects in high-speed relative motion, but what we think we will see given the two postulates of special relativity.

In classical mechanics, we usually have at least three main "actors": two reference frames and a body whose kinematic properties are under analysis. Reference frame transformations have thus a sense in order to give a consistent picture of the same phenomenon (kinematic properties of the third body) in the two different reference frames. In the case of light pulses, however, there is no longer the "same object" seen by different reference frames, and thus kinematic transformations in that case may have a different meaning and utility.

As a matter of fact, almost all the derivations of the special relativity consequences (starting from the relativity of simultaneity up to Lorentz transformations) made by Einstein in his 1905 paper, and almost all the thought experiments on special and general relativity in the following developments (e.g. the thought experiment of the deflection of a beam of light in a free-falling elevator), consider a beam of light as a "third body" that maintains its own identity[11] from a reference frame to the other, much like a physical

---

[11]Consider, further, the oft-quoted "chasing the light" thought experiment devised by Einstein: in that thought experiment, he considers the beam of light as an object that maintains its own identity when it is seen from different

object (e.g. a ball).

The second postulate may be rewritten as follows: light travels at the same speed c in every reference frame, but this does not mean that we are talking about the *same light* (same object). The concept of "same light" may have no meaning at all. There should be no longer need for the Lorentz transformations in order to force a beam of light (wrongly intended as an independent physical object maintaining its own identity) to have the same velocity in two relatively moving inertial reference frames. After all, the by-now widely held belief in the non-existence of a light-bearing "material" medium hints exactly at this. For instance, a sea wave is a unique and independent physical object because there is a material medium through which it propagates; but, as far as we know, a beam of light does not propagate through any "material" medium. Thus, we ask, why should it be taken for granted that a light beam must behave as an independent physical object maintaining its own identity when observed from two relatively moving inertial reference frames?

We acknowledge that the reflections above are no more than a sketch lacking any rigorous mathematical framework; but a light beam seen from two different reference frames is absolutely not the *same thing*, as happens in the case of a material body, and events defined by the emission and absorption of light beams in a reference system may be no longer *related* to these beams of light when the beams and the events are observed from a different reference system; thus, it does not make any sense at all to talk about simultaneity (or lack of simultaneity). In this sense, Lorentz transformations are a mere mathematical expedient to bind distinct physical events, so to speak. No one really knows how asynchronous the events happening in the world around us would appear if ever we could see them by traveling at a speed close to that of light. Moreover, such a lack of synchronicity is, according to our main contention, only apparent and does not leave any permanent trace like asymmetric ageing.

A final remark, if we accept for a moment the conclusions above

---

inertial frames.

as consistent with physical reality, we cannot fail to notice that a worrying shadow is being cast over the theoretical results of general relativity too, at least in the way in which they were initially derived by Einstein. If Lorentz transformations do not describe actual, physical time dilation and length contraction, then even the gravitational time dilation first derived by Einstein in 1907 [5] by using special relativity and the principle of equivalence cannot be an actual, physical time dilation. According to an early version of the principle of equivalence, uniformly accelerated frames are equivalent to frames at rest in a suitable gravitational field and vice-versa. By only applying the machinery of special relativity, Einstein derived gravitational time dilation (intended to be actual, physical) from Lorentz transformations applied from within an inertial reference frame to a uniformly accelerated frame (this frame representing a system in a gravitational field[12]). How can real, physical time dilation (gravitational time dilation) come from an apparent effect (special relativity time dilation coming from Lorentz transformations)?

As a matter of fact, Einstein gave a simpler derivation of gravitational time dilation in a paper published in 1911 [6], where he derived the *gravitational redshift* by means of the Doppler effect of light falling from the top (S_2) to the bottom (S_1) of an upwardly accelerating elevator. If acceleration is assumed to be equivalent to a gravitational field, the same result holds if the system is stationary in a uniform gravitational field. Einstein then argues that this result seems to assert an absurdity: if there is constant transmission of light from S_2 to S_1, how can any other number of periods per second (frequency) arrive at S_1 than is emitted from S_2? His resolution of this apparent paradox is that "nothing compels us to assume that the clocks in different gravitation potentials must be regarded as going at the same rate". Namely, the number of periods is the same but the time rate in two positions with different gravitation potential is not. By following the same line of thought

---

[12]According to the principle of equivalence, in this derivation, the inertial frame stands for a frame free falling in a gravitational field. Thus, we must conclude that Einstein saw no trouble in applying special relativity from within a free falling frame; see Subsection 7.1.1.

described in this section, in this case, we might say that the light pulse emitted in S_2 is not the *same thing* as that observed in S_1; thus, why bother about different numbers of periods per second?

7.3   Voices of dissent

It is interesting to note that at least one distinguished scholar (actually, more than one) has hitherto reached analogous conclusions about the reality of time dilation in special and general relativity, although by following a different path. In a series of papers published in the 1970s and in the 1980s [28, 30, 31, 32, 42, 46], Mendel Sachs clearly stated that:

> [...] on the basis of a rigorous mathematical formulation of the theory of general relativity (whose results incorporate those of special relativity), as well as an analysis of the logical structure of the theory, *there is no prediction of asymmetric ageing* and there is no logical paradox in the physical prediction of the theory. I believe that Einstein's identification of the Lorentz transformation with a physical cause-effect relation, and the subsequent conclusion about asymmetric ageing, *was a flaw*, not in the theory of relativity itself, [...], but rather *a flaw in the reasoning that Einstein used in this particular study*–leading him to an inconsistency with the meaning of space and time, according to his own theory. [32] [emphasis added]

and,

> [...] the theory of relativity, *per se*, does not imply an actual asymmetric physical effect on the ageing of a physical entity ("observed") that is in motion relative to any other physical entity ("observer"). [42]

Moreover, he said that:

[...] the crux of my argument was that the essence of Einstein's theory implies that the space-time transformations between relatively moving frames of reference must be interpreted strictly kinematically, rather than dynamically. Thus, according to this theory, the transformations are not more that necessary scale changes that must be applied to the *measures* of space and time [...] [42]

It is impossible to believe that Sachs, at least at the time of his last publications on the topic, was not aware of some the most famous experimental "confirmations" of time dilation (e.g. muon decay, Hafele and Keating's experiment and GPS synchronization, to name a few). Therefore, by taking his mathematical and logical arguments at their face value, he should have come to the same conclusion as ours: the alleged experimental confirmations of physical time dilation (asymmetric ageing) are in fact at odds with the theory of special relativity in its original formulation or, at least, they are physically and logically unrelated to it.

7.4 Appendix A: More on Purcell's basic explanation of magnetic forces

It is argued (see, for instance, Born [22]) that the magnetic force felt by a charge moving nearby a fixed wire passed through by a current is in fact an electrostatic force when seen in the rest frame of the moving charge. This is explained by the fact that in the rest frame of the moving charge the lattice positive charges of the wire are seen as moving and then their density is augmented by Lorentz contraction in a different way than the moving electrons in the wire which constitute the current (different Lorentz contraction). This charge unbalance, *which must be in fact real and physical and not only apparent*, is allegedly at the origin of the force experienced by the moving charge (Lorentz force). This explanation appears to be one of the most striking successes of special relativity: the calculations fit perfectly with the phenomenological laws of magnetic forces. Moreover, it seems to be one of the few relativity consequences affecting physical phenomena close to everyday life (and it is, in some sense, a stunning "mechanical" explanation of magnetism).

There is one thing, however, that appears to be not so crystal-clear. Inside a metal wire at rest in the laboratory and passed through by a current, only the electrons move (drift). To an observer at rest, their density should appear augmented with respect to the density of the stationary lattice positive charges (which is the same density of the electrons when the current is off) by a suitable Lorentz factor. Instead, in every quantitative treatment of the problem we have seen, even when the current is on, moving electrons charge density in the frame of the wire is considered to be equal to the charge density of the stationary lattice positive charges and, to all accounts, the wire is consistently treated as to be neutral in the laboratory frame. Unless we are ready to accept that no Lorentz contraction takes place on the moving electrons with respect to the laboratory frame, we have to explain this behavior as though, in the laboratory frame, the electrons with a Lorentz augmented density rearrange themselves in order to give a zero net charge on the wire, more or less according to the same principle by which charges spread over a metallic surface in order to reduce the overall electric potential difference. So: why do they not do the same to reduce the overall electric potential difference (zero net charge) when they are seen in the reference frame of the charge moving near the wire? We see in all this a lack of reciprocity, the same reciprocity demanded by the principle of relativity. If the relativistic "excess" of positive charge on the wire, when seen by the charge moving nearby the wire, is real enough to attract that charge, then this same excess of positive charge should be real enough (again, when seen in the reference frame of the moving charge) to attract more electrons along the wire from the device that generates the current across that wire in order to reestablish a zero net charge. Why does it not happen?

There are few doubts that Purcell's relativistic explanation is quite impressive. However, can this same relativistic approach explain Ampère's force law between two parallel wires passed through by a current in the same direction? It probably explains why an electron moving in one of the wires sees an attractive force exerted by the other wire. If we consider for a moment what an "external observer" at rest in the same reference frame of the electron sees (the observer moves at the same velocity of the electron current), things become more difficult. According to this observer, *both* wires become overall positively charged (we simply apply the contraction argument to both wires) and they should repel each other in her frame, exactly the contrary of what Ampère's force law and experience tell us. According to the principle of relativity, in this reference frame, Ampère's force law must hold true; in fact in this frame, the observer sees two parallel currents (this time due to the lattice positive charges seen as moving in the opposite direction). Yet, in order to have the same net attractive force as that measured in the laboratory frame, in this reference frame the wires must be overall neutral: is this the charge rearrangement argument of which we have just talked? Two considerations follow in order: if the charge rearrangement argument also holds true in the reference frame of the moving charge, then the relativistic explanation of the force experienced by a single charge moving nearby the wire vanishes (but then, how to explain the Lorentz force acting upon the charge? In the reference frame where the charge is at rest, no force should act upon it). Moreover, if we need to resort to Ampère's force law to explain what happens between two wires passed through by a current when viewed from a reference frame moving at the same velocity of the charges of the current, then the relativistic explanation of the magnetic force is superfluous.

Finally, consider the following thought experiment. Take an infinite solenoid passed through by a current. From Maxwell's equations, it is known that there is no magnetic field outside the solenoid (and, obviously, no electric field). However, if one sets the solenoid in rotation about its main axis, according to the very same relativity explanation of magnetic forces, an observer at rest outside will now see the solenoid charging up, positively or negatively according to the direction of the current and the direction of rotation. Alternatively, an observer rotating around the fixed solenoid will feel an electrostatic field. All this appears to be at odds with experimental observations: no charge, moving or at rest, is affected by any macroscopic force whatsoever outside (and close to) a sufficiently long solenoid passed through by a current. Note that the force to which we refer must not be confused with the magnetic force due to the external and weaker magnetic field that should be present in the more realistic case of a non-infinite solenoid. This last force can be neglected if the solenoid is suitably long and/or if we study the forces close enough to the solenoid from the outside. On the contrary, the intensity of the alleged electrostatic force due to special relativity can reasonably be made as high as we want, by simply acting on the solenoid rotation velocity.

7.5 Appendix B: Upper limit on speeds

Nothing material can overcome or even reach the speed of light. This appears to be the inescapable consequence of special relativity. Einstein elaborated on this in his 1907 paper [5] where he applied what he called the "addition theorem of velocities" to show that there cannot exist an effect that can be used for arbitrary signaling and that is propagated faster than light in a vacuum.

As a matter of fact, as far as the upper limit on speeds is concerned, we believe that it was already in the postulates of special relativity. It is actually a case of *petitio principii*. Within special relativity, the Lorentz transformations were derived by considering a beam of light as a physical object maintaining its own identity and by imposing that this *object* has the same velocity c in every reference frame. In this way, such transformations have already been built to mathematically force objects (every objects) to move at speeds not higher that the speed of light.

If we adopt the idea proposed in the previous sections, namely that it is true that light travels always at speed c, but that a beam of light seen from two different inertial reference frames is not the same beam as would happen with a material object, then we do not need Lorentz transformations to "force" a light beam to have the same velocity in every reference frame. In this way, the constancy of the velocity of light would no longer imply, via Lorentz transformations, an upper limit on the speed of material objects.

Interestingly enough, Van Flandern collected a series of not-so-easily dismissible pieces of evidence [61] that, for example, the "speed of gravity" is actually greater than the speed of light. For instance, the path of planetary bodies (asteroids, comets and even "small" planets) seems to be extremely well predicted if one considers the gravitational force exerted at the *same instant of time* by all major planets from their current positions, notwithstanding their different distances from the body under study and thus, without taking into account the different delays with which the body feels the actual positions of all planets. This delay would necessarily be present if gravity had a finite speed (or, at least, a speed not greater than c). How can this be possible? In the long run, this delay should show up and should outlast any other non-gravitational or relativistic perturbations on the motion of celestial bodies.

7.6 Appendix B: Reflections upon the derivation of the mass-energy equivalence

Let us consider the elementary derivation of the mass-energy formula $E = mc^2$ proposed by Einstein in 1946 [16] which, according to his own words, although it makes use of the principle of special relativity, does not presume the formal machinery of the theory but uses only three previously known laws:

(1) The law of the conservation of momentum;

(2) The expression for the pressure of radiation; that is, the momentum of a complex of radiation moving in a fixed direction;

(3) The well known expression for the aberration of light (i.e the influence of the motion of the Earth on the apparent location of the fixed stars–Bradley).

As a matter of fact, we believe that the following observations can also be applied to the original and subsequent derivations by other authors. In the derivation, Einstein describes the absorption of two complexes of radiation by a body of mass M as seen from two reference inertial frames: K_0, the rest frame of the body, and K, a frame moving with respect to K_0 with velocity v.

In the rest frame K_0, the two complexes of radiation move towards body B in opposite directions and thus no net linear momentum is transferred to B, and the body stays at rest even after the absorption. In the moving frame K, instead, the two complexes of radiation are seen as having different directions (due to light aberration) and are no longer strictly "opposing": this means that, according to the reference frame K, after the absorption body B has a higher linear momentum which can only be ascribable to an increase of mass, since its velocity after the absorption in the frame K_0 does not change and so it must be also with respect to frame K. According to the calculations and the three laws cited before, this mass increase turns out to be equal to the whole energy E of the two complexes of radiation divided by c^2.

However, according to the idea proposed in the previous sections, if a beam of light observed from two different inertial frames is not to be intended as the same beam, the two complexes of radiation seen in the reference frame K are not the same as those actually absorbed by body B in its reference frame K_0. Therefore, there is no need to appeal to a mass increase to explain an increase in the linear momentum of body B, because there is actually no increase in the linear momentum at all. In some sense, the two complexes of radiation seen by an observer in K (and typical of this reference frame) can only be absorbed by bodies stationary in K and not by bodies moving in K_0, more or less as it happens in the permanent magnet thought experiment described in Section 7.2, where radiation generated in one reference frame can be detected and absorbed only by bodies in that reference frame, while in another reference frame (e.g. the magnet rest frame) the radiation is even non-existent.

In the original derivation of 1905, Einstein did actually make implicit use of the law of momentum conservation. In fact, he mainly performed an energy balance calculation on a body emitting planes waves of light in opposite directions. According to special relativity, the energies of the waves of light are different when seen from a reference frame moving with respect to the rest frame of the emitting body. Einstein ascribed this difference of energy to the variation of the kinetic energy of the body but, since the emission of plane waves had been in opposing directions in the rest frame and this could not change the velocity of the body, the difference of energy must necessarily come from a change in the mass of the body. Once more, if we want to read this derivation in light of the idea proposed in the previous sections, here Einstein conflates two physical situations that cannot be mixed up: the plane waves of light emitted in the rest frame of the body and those observed in the moving frame are not the same physical object and there is no clear reason why the energy difference must be attributed to the physical properties of the body. Again, by referring to the permanent magnet thought experiment described in Section 7.2, we noticed that in the moving frame, em waves originate that are not detectable in the reference frame of the permanent magnet and are not attributable to any physical property the magnet itself: the magnet does not accelerate; it does not emit anything; and its mass is constant. Thus, Maxwell's theory of light warns us that when electromagnetic phenomena are considered from different inertial frames, we must be careful with energy balance evaluation. In the classical theory of light, there is an energy balance conundrum that must be accounted for and cannot be resolved by the equivalence $E = mc^2$. The energy difference accounted for in the formula $E = mc^2$, and ascribed to the mass variation of the emitting body, appears to be actually related, at least in the way in which it has been originally derived, to the em field and, more generally, to the physics described by Maxwell's equations (as in the case of the permanent magnet thought experiment).

Chapter 8

Outlook

At first sight, the two topics covered in this book (viz. thermionic/photoelectric phenomena and the second law on the one hand; special relativity on the other) may appear too distant and esoteric to some readers. They certainly were when they first appeared as distinct matters of investigation in the mind of the author. Obviously, we managed to find a faint *fil rouge* in the title. In this book, we have probed two well known and widely held limits of Physics: that posed by the second law of thermodynamics and that posed by special relativity (which sets the speed of light as the upper limit for the speeds of objects with positive rest mass).

However, retrospectively, we realized that a deeper *fil rouge* connects the two topics, apparently by accident, which is in the physical phenomenon of light and in the laws that govern its generation, propagation and interaction with matter. The connection between thermionic/photoelectric phenomena and the second law stands mainly on the laws that govern black-body radiation and black-body radiation is light that interacts with matter in thermal equilibrium. Furthermore, the theory of special relativity would not have existed without Maxwell's theory of light and equations, and the foundational questions they raised in relation to the principle of relativity.

The more we reflect on Maxwell's theory of light and Maxwell's equations, the more we believe that the questions they posed, and that have been allegedly solved by the theory of special relativity, are actually still present. In our opinion, more than the lofty "space-time" jargon of relativity, are Maxwell's equations and Maxwell's theory the most baffling and counterintuitive part of Physics as a whole. Light is still a mystery; we believe that there is much more yet to discover.

Bibliography

[1] Maxwell, J.C. (6 Dec 1870) "Letter to John William Strutt", in *The Scientific Letters and Papers of James Clerk Maxwell (1995)*, P. M. Hannan, (Ed.), Vol. 2, 582–583, Cambridge University Press, Cambridge, UK.

[2] Maxwell, J.C. (1871) *Theory of Heat*, Longmans, Green, and Co., London.

[3] Original German version: Einstein, Albert (1905), "Zur Elektrodynamik bewegter Körper", *Annalen der Physik* 322(10): 891921.
Translated version: https://en.wikisource.org/wiki/On_the_Electrodynamics_of_Moving_Bodies_(1920_edition).

[4] Poincaré, H. (1905) "The Principles of Mathematical Physics", *The Monist*, 15(1), 1–24.

[5] Einstein, Albert (1907) "Über das Relativitätsprinzip und die aus demselben gezogenen Folgerungen", *Jahrbuch der Radioaktivität und Elektronik* 4, 411462.

[6] Einstein, Albert (1911) "Einfluss der Schwerkraft auf die Ausbreitung des Lichtes", *Annalen der Physik* 35, 898–908.

[7] Smoluchowski, M. von, (1914) "Gültigkeitsgrenzen des zweiyen Hauptsatzes der Wärmtheorie", *Vorträge über die Kinetische Theorie der Materie der Elektrizität*, 89–121, Teubner, Leipzig.

[8] Einstein, A. (1916) *Relativity: The Special and General Theory*, Methuen & Co Ltd, London, 1920.

[9] Einstein, A. (1918) "Dialog about Objections against the Theory of Relativity", *Die Naturwissenschaften* 6 (48), 697-702.

[10] Eddington, A.S. (1929) *The Nature of the Physical World*, Cambridge University Press, Cambridge, p. 74.

[11] O. W. Richardson's Nobel lecture on thermionics, 1929, 224-236. Available at: http://www.nobelprize.org/nobel_prizes/physics/laureates/1928/richardson-lecture.pdf (April 2015)

[12] Szilard, L. (1929), "On the Decrease of Entropy in a Thermodynamic System by the Intervention of Intelligent Beings", *Zeitschrift fur Physik*, 53, 840-856.

[13] Sommer, A.H. (1936) *Photoemissive materials: preparation, properties, and uses*, Section 7.1, Chapter 10, John Wiley & Sons.

[14] Gimpel, I., Richardson, O. (1943) "The Secondary Electron Emission from Metals in the Low Primary Energy Region", *Proceedings of the Royal Society of London A*, 182, 17-47.

[15] Brillouin, J., (1950) "Can the Rectifier Become a Thermodynamical Demon?", *Physics Reviews*, 78(5), 627-628.

[16] Einstein, A. (1950) *Out of My Later Years: The Scientist, Philosopher, and Man Portrayed Through His Own Words*, Open Road, New York.

[17] Brillouin, L. (1951) "Maxwell's Demon cannot operate: Information and entropy I", *Journal of Applied Physics*, 22, 334-337.

[18] Sommerfeld, A. (1952) *Vorlesungen über theoretische Physik, Thermody-namik und Statistik*, Band 5, Dieterich'sche Verlagsbuchhandlung, Wiesbaden.

[19] Fermi, E. (1956) *Thermodynamics*, Dover Publications, Inc., New York.

[20] Sommer, A.H. (1956) "Multi-Alkali Photo Cathode", *IRE Transactions on Nuclear Science*, 8–12. Invited paper presented at Scintillation Counter Symposium, Washington, D.C., 1956.

[21] Landauer, R. (1961) "Irreversibility and heat generation in the computing process", *IBM Journal of Research and Development*, 5, 183–191.

[22] Born, M. (1962) *Einstein's Theory of Relativity*, Dover Publications Inc.

[23] Feynman, R.P., Leigthon, R.B., Sands, M. (1963) *The Feynman Lectures on Physics*, Vol. I, Addison-Wesley Publishing Company.

[24] Rohrlich, F. (1965) *Classical Charged Particles. Reading*, Addison-Wesley.

[25] Dannhäuser, F. (1967) "Analysis of bulk reverse current in diffused silicon power rectifiers", *Solid State Electronics*, 10, 361–365.

[26] Uebbing, J.J., James, L.W. (1970) "Behavior of Cesium Oxide as a Low WorkFunction Coating", *Journal of Applied Physics*, **41**(11), 4505

[27] McFee, R. (1971) "Self-Rectification in Diodes and the Second Law of Thermodynamics", *American Journal of Physics*, 39, 814.

[28] Sachs, M. (1971) "A resolution of the clock paradox", *Physics Today*, 24(9), 23–29.

[29] Holstein, B.R. and Swift, A.R. (1972) "The Relativity Twins in Free Fall" *Am. J. Phys.*, 40, 746.

[30] Terrell, J. *et al.* (1972) "The clock "paradox"— majority view", *Physics Today*, 25(1), 9.

[31] Sachs, M. (1973) "Comments on the clock paradox", *International Journal of Theoretical Physics*, 7(4), 281–285.

[32] Sachs, M. (1974) "On Dingle's Controversy about the Clock Paradox and the Evolution of Ideas in Science", *International Journal of Theoretical Physics*, 10(5), 321–331.

[33] Lerner, A.Y. (1975) *Fundamentals of Cybernetics*, Plenum Pub. Corp., New York, 257.

[34] Boulware, D.G. (1980) "Radiation from a Uniformly Accelerated Charge", *Annals of Physics.*, 24, 169–188.

[35] Bates Jr. C.W. (1981) "Photoemission from Ag–O–Cs", *Physical Review Letters*, 47(3), 204–208.

[36] Denur, J. (1981) "The Doppler demon", *American Journal of Physics*, 49, 352–355.

[37] Gordon, L.G.M. (1981) "Brownian movement and microscopic irreversibility", *Foundations of Physics*, 11, 103–113.

[38] Bennett, C.H. (1982) "The thermodynamics of computation– a review", *International Journal of Theoretical Physics*, 5(12), 905–940.

[39] Gordon, L.G.M. (1983) "Maxwell's demon and detailed balancing", *Foundations of Physics*, 13, 989–997.

[40] Motz, H. (1983) "The Doppler demon exorcised", *American Journal of Physics*, 51, 71–72.

[41] Chardin, G. (1984) "No free lunch for the Doppler demon", *American Journal of Physics*, 52, 252–253.

[42] Sachs, M. (1985) "On Einstein's later view of the twin paradox", *Foundations of Physics*, 15(9), 977–980.

[43] Bennett, C.H. (1987) "Demons, Engines and the Second Law", *Scientific American*, 257(5), 108–116.

[44] Rossi, D. V., Fossum, E. R., Pettit, G. D., Kirchner, P. D., Woodall, J. M. (1987) "Reduced reverse bias current in AlGaAs and $In_{0.75}Ga_{0.25}AsGaAs$ junctions containing an interfacial arsenic layer", *Journal of Vacuum Science & Technology B*, 5, 982.

[45] Denur, J. (1989) "Velocity-dependent fluctuations: Breaking the randomness of Brownian motion", *Physical Review A*, 40, 5390–5399.

[46] Sachs, M. (1989) "Response to Rodrigues and Rosa on the twin paradox", *Foundations of Physics*, 19(12), 1525–1528.

[47] Gordon, L.G.M. (1994) "The molecular-kinetic theory and the second law", *Journal of Colloid and Interface Science*, 162, 512–513.

[48] Cotti, P. (1995) "The discovery of the electric current", *Physica B: Condensed Matter*, 204, 367–369.

[49] Garber, E., Brush, S.G., Everitt, C.W.F. (1995) *Maxwell on Heat and Statistical Mechanics. On "Avoiding All Personal Enquiries"*. Associated University Presses, London.

[50] Sheehan, D.P. (1995) "A paradox involving the second law of thermodynamics", *Physics of Plasmas*, 2, 1893–1898.

[51] Crosignani, B., Di Porto, P. (1996) "Approach to thermal equilibrium in a system with adiabatic constraints", *American Journal of Physics*, 64, 610–613.

[52] Sheehan, D.P. (1996) "Another paradox involving the second law of thermodynamics", *Physics of Plasmas*, 3, 104–110.

[53] Čápek, V. (1997a) "Isothermal Maxwell daemon and active binding pairs of particles", *Journal of Physics A: Mathematical and General*, 30, 5245–5258.

[54] Čápek, V. (1997b) "Isothermal Maxwell daemon", *Czechoslovak Journal of Physics*, Vol. 47, 845-849.

[55] Earman, J., Norton, J.D. (1998) "Exorcist XIV: The Wrath of Maxwell's Demon. Part I. From Maxwell to Szilard", *Studies in the History and Philosophy of Modern Physics*, 29(4), 435–471.

[56] Nikulov, A. V., Zhilyaev, I. N. (1998) "The Little-Parkes effect in an inhomogeneous superconducting ring", *Journal of Low Temperature Physics*, 112, 227–236.

[57] Sheehan, D.P., Means, J.D. (1998) "Minimum requirements for second law violation: A paradox revisited", *Physics of Plasmas*, 5, 2469–2471.

[58] Sheehan, D.P. (1998) "Dynamically maintained steady-state pressure gradients", *Physical Review E*, 57, 6660–6666.

[59] Čápek, V. (1998) "Isothermal Maxwell daemon: Swing (fish-trap) model for particle pumping", *Czechoslovak Journal of Physics*, 48, 879–901.

[60] Čápek, V., Bok, J. (1998) "Isothermal Maxwell daemon: Numerical results in a simplified model", *Journal of Physics A: Mathematical and General*, 31, 8745–8756.

[61] Van Flandern, T. (1998) "The speed of gravity – What the experiments say", *Physics Letters A*, 250, 1–11.

[62] Van Flandern, T. (1998) "What the Global Positioning System Tells Us about Relativity", in *Open Question in Relativistic Physics*, F. Selleri, ed. pp. 81–90, Apeiron, Montreal.

[63] Čápek, V., Mančal, T. (1999) "Isothermal Maxwell daemon as a molecular rectifier", *Europhysics Letters*, 48, 365–371.

[64] Čápek, V., Bok, J. (1999) "A thought construction of working perpetuum mobile of the second kind", *Czechoslovak Journal of Physics*, 49, 1645–1652.

[65] Čápek, V., Tributsch, H. J. (1999) "Particle (electron, proton) transfer and self-organization in active thermodynamic

reservoirs", *Journal of Physical Chemistry B*, 103, 3711–3719.

[66] Earman, J., Norton, J.D. (1999) "Exorcist XIV: The Wrath of Maxwell's Demon. Part II. From Szilard to Landauer and Beyond", *Studies in the History and Philosophy of Modern Physics*, 30(1), 1–40.

[67] Allahverdyan, A. E., Nieuwenhuizen, Th. M. (2000) "Extraction of work from a single thermal bath in the quantum regime", *Physical Review Letters*, 85, 1799–1802.

[68] Čápek, V., Frege, O. (2000) "Dynamical trapping of particles as a challenge to statistical thermodynamics", *Czechoslovak Journal of Physics*, 50, 405–423.

[69] Sheehan, D.P. (2000) Reply to "Comment on 'Dynamically maintained steady-state pressure gradients.' ", *Physical Review E*, 61, 4662–4665.

[70] Sheehan, D.P., Glick, J. Means, J.D. (2000) "Steady-state work by an asymmetrically inelastic gravitator in a gas: A second law paradox", *Foundations of Physics*, 30, 1227–1256.

[71] Sheehan, D.P., Glick, J. (2000) "Gravitationally-induced, dynamically-maintained, steady-state pressure gradients", *Physica Scripta*, 61, 635–640.

[72] Allahverdyan, A. E., Nieuwenhuizen, Th. M. (2001) "Breakdown of the Landauer bound for information erasure in the quantum regime", *Physical Review E*, 64, 056117.

[73] Čápek, V., Bok, J. (2001) "Violation of the second law of thermodynamics in the quantum microworld", *Physica A*, 290, 379–401.

[74] Čápek, V. (2001) "Twilight of a dogma of statistical thermodynamics", *Molecular Crystals and Liquid Crystals*, 335, 13–24.

[75] Crosignani, B., Di Porto, P. (2001) "On the validity of the second law of thermodynamics in the mesoscopic realm", *Europhysics Letters*, 53, 290–296.

[76] Hsu, J. W. P., Manfra, M. J., Lang, D. V., Richter, S., Chu, S. N. G., Sergent, A. M., Kleiman, R. N., Pfeiffer, L. N., Molnar, R. J. (2001) "Inhomogeneous spatial distribution of reverse bias leakage in GaN Schottky diodes", *Applied Physics Letters*, 78, 1685.

[77] Nikulov, A. V. (2001) "Quantum force in a superconductor", *Physical Review B*, 64, 012505.

[78] Sheehan, D.P. (2001) "The second law and chemical induced, steady-state pressure gradient: controversy, corroboration and caveats", *Physica A*, 280, 185–190.

[79] Uffink, J. (2001) "Bluff your way in the second law of thermodynamics", *Studies in the History and Philosophy of Modern Physics*, 32, 305–394

[80] Čápek, V., Bok, J. (2002) "A model of spontaneous generation of population inversion in open quantum systems", *Chemical Physics*, 277, 131–143.

[81] Čápek, V., Mančal, T. (2002) "Phonon mode cooperating with particle serving as Maxwell gate and rectifier", *Journal of Physics A: Mathematical and General*, 35, 2111–2130.

[82] Čápek, V., Sheehan, D.P. (2002) "Quantum mechanical model of a plasma system: A challenge to the second law of thermodynamics", *Physica A*, 304, 461–479.

[83] Čápek, V. (2002) "Zeroth and second law of thermodynamics simultaneously questioned in the quantum world", *European Physical Journal B*, 25, 101–113.

[84] Duncan, T.L. (2002) in *Quantum Limits to the Second Law: First International Conference*, D.P. Sheehan, Editor, AIP Conference Proceedings, 643, AIP Press, Melville, NY.

[85] Nieuwenhuizen, Th. M., Allahverdyan, A.E. (2002) "Statistical thermodynamics of quantum Brownian motion: Construction of perpetuum mobile of the second kind" *Physical Review E*, 66, 036102.

[86] Oyama, S., Hashizume, T., Hasegawa, H. (2002) "Mechanism of current leakage through metal/n-GaN interfaces", *Applied Surface Science*, 190, 322–325.

[87] D. P. Sheehan (Ed.) (2002) *First International Conference on Quantum Limits to the Second Law*, AIP Conference Proceedings, 643, AIP Press, Melville, NY.

[88] Sheehan, D.P., Glick, J., Duncan, T., Langton, J.A., Gagliardi, M.J., Tobe, R. (2002a) "Phase space portraits of an unresolved gravitational Maxwell demon", *Foundations of Physics*, 32, 441–462.

[89] Sheehan, D.P., Glick, J., Duncan, T., Langton, J.A., Gagliardi, M.J., Tobe, R. (2002b) "Phase space analysis of a gravitationally-induced, steady-state nonequilibrium", *Physica Scripta*, 65, 430–437.

[90] Sheehan, D.P., Wright, J.H., Putnam, A.R. (2002c) "A solid-state Maxwell demon", *Foundations of Physics*, 32, 1557–1595.

[91] Allahverdyan, A. E., Nieuwenhuizen, Th. M. (2003) "Testing the violation of the Clausius inequality in nanoscale electric circuits", *Physical Review B*, 66, 115–309.

[92] Bennett, C.H. (2003) "Notes on Landauer's principle, reversible computation, and Maxwell's Demon", *Studies in the History and Philosophy of Modern Physics*, 34, 501–510.

[93] Čápek, V. (2003) "Dimer as challenge to the second law", *European Physical Journal B*, 34, 219–223.

[94] Keefe, P. (2003) "Coherent magneto-caloric effect superconductive heat engine process cycle", *Journal of Modern Optics*, 50, 2443–2454.

[95] Klein, U., Vollmann, W., Abatti, P.J. (2003) "Contact potential differences measurements: Short history and experimental setup for classroom demonstration", *IEEE Trans. Educ.*, 46, 338–344.

[96] Leff, H.S., Rex, A.F. (2003) *Maxwell's Demon 2: Entropy, Information Computing.* Institute of Physics, Bristol.

[97] Allahverdyan, A. E., Nieuwenhuizen, Th. M. (2004) "Bath-generated work extraction and inversion-free gain in two-level systems", *Journal of Physics A: Mathematical and General*, 36, 875–882.

[98] Bok, J., Čápek, V. (2004) "Langevin approach to the Porto system", *Entropy*, 6, 5767.

[99] Callender, C. (2004) "A collision between Dynamics and Thermodynamics", *Entropy*, 6, 11–20.

[100] Crosignani, B., Di Porto, P., Conti, C. (2004) "The adiabatic piston: A perpetuum mobile in the mesoscopic realm", *Entropy*, 6, 50–56.

[101] Denur, J. (2004) "Modified Feynman ratchet with velocity-dependent fluctuations", *Entropy*, 6, 76–86.

[102] Gijsbers, V. (2004) *The contingent law: A tale of Maxwell's Demon* (philsci-archive.pitt.edu/2201/)

[103] Gordon, L.G.M. (2004) "The decrease of entropy via fluctuations", *Entropy*, 6, 38–49

[104] Gordon, L.G.M. (2004) "A Maxwellian valve based on centrifugal forces", *Entropy*, 6, 87–95.

[105] Gordon, L.G.M. (2004) "Smoluchowski's trapdoor", *Entropy*, 6, 96–101.

[106] Keefe, P. (2004) "Second law violation by magneto-caloric effect adiabatic phase transition of type I superconductor particles", *Entropy*, 6, 116–127.

[107] Keithley Instruments. Low Level Measurements Handbook: Precision DC Current, Voltage, and Resistance Measurements (6th edition: 2004).

[108] Wheeler, J.C. (2004) "Resolution of a Classical Gravitational Second-Law Paradox", *Foundations of Physics*, 34(7), 1029–1062.

[109] Abiko, S. (2005) "The Origins and Concepts of Special Relativity", in *Physics Before and After Einstein* M. Mamone Capria (Ed.) pp. 71–91, IOS Press, Amsterdam.

[110] Berger, J. (2005) "The Chernogolovka experiment", *Physica E*, 29, 100–103.

[111] Čápek, V., Sheehan, D. P. (2005) *Challenges to the Second Law of Thermodynamics–Theory and Experiments, Fundamental Theories of Physics*, 146, Springer, Dordrecht, Netherlands.

[112] Gyftopoulos, E.P., Beretta, G.P. (2005) "What is the second law of thermodynamics and are there any limits to its validity?" http://arxiv.org/abs/quant-ph/0507187 (June 2012)

[113] Keefe, P. (2005) "Quantum limit to the second law by magneto-caloric effect, adiabatic phase transition of mesoscopic-size type I superconductor particles", *Physica E*, 29, 104–110.

[114] Norton, J.D. (2005) "Eaters of the Lotus: Landauer's Principle and the Return of Maxwell's Demon" *Studies in the History and Philosophy of Modern Physics*, 36, 375–411.

[115] Sheehan, D.P., Seideman, T. (2005) "Intrinsically-biased electrocapacitive catalysis", *Journal of Chemical Physics*, 122, 204713.

[116] Sheehan, D.P., Wright, J.H., Putnam, A.R., Pertuu, A.K. (2005) "Intrinsically-biased, resonant NEMS-MEMS os-

cillator and the second law of thermodynamics", *Physica E*, 29, 87–99.

[117] Unnikrishnan, C. S. (2005), "On Einsteins resolution of the twin clock paradox", *Current Science* 89(12), 2009–2015.

[118] de Almeida, C. and Saa, A. (2006) "The radiation of a uniformly accelerated charge is beyond the horizon: A simple derivation", *American Journal of Physics*, 74, 154.

[119] Sheehan, D.P., Gross, D.H.E. (2006) "Extensivity and the thermodynamic limit: Why size really does matter", *Physica A*, 370, 461–482.

[120] Sheehan, D.P. (Ed.) (2006) *The Second Law of Thermodynamics: Foundation and Status*, Proceedings of Symposium at 87th Annual Meeting of the Pacific Division of AAAS, University of San Diego.

[121] Berger, J. (2007) "A nonconventional scenario for thermal equilibrium", *Foundations of Physics*, 37, 1738–1743.

[122] Crosignani, B., Di Porto, P. (2007) "Random fluctuations of diathermal and adiabatic pistons", *Foundations of Physics*, 37, 1707–1715.

[123] Denur, J. (2007) "Speed-dependent weighting of the Maxwellian distribution in rarefied gasses: A second law paradox?", *Foundations of Physics*, Vol. 37, 1685-1706.

[124] Sheehan, D.P. (2008) "Energy, Entropy and the Environment (How to Increase the First by Decreasing the Second to Save the Third)", *Journal of Scientific Exploration*, 22(4), 459–480.

[125] Maroney, O.J.E. (2009) *Information Processing and Thermodynamic Entropy*, Stanford Encyclopedia of Philosophy (plato.stanford.edu/entries/information-entropy/)

[126] D'Abramo, G. (2010) "Thermo-charged capacitors and the Second Law of Thermodynamics", *Physics Letter A*, 374, 1801–1805.

[127] D'Abramo, G. (2011) "On the exploitability of thermo-charged capacitors", *Physica A*, 390, 482–491. Available at http://arXiv.org/abs/0912.4818 [physics.gen-ph].

[128] D'Abramo, G. (2011) Addendum to "on the exploitability of thermo-charged capacitors". *Physica A*, 391, 482–491. Available at http://arXiv.org/abs/0912.4818 [physics.gen-ph].

[129] Norton, J.D. (2011) "Waiting for Landauer", *Studies in the History and Philosophy of Modern Physics*, 42(3), 184–198.

[130] Perminov, A., Nikulov, A. (2011) "Transformation of Thermal Energy into Electric Energy via Thermionic Emission of Electrons from Dielectric Surfaces in Magnetic Fields", in *Second Law of Thermodynamics: Status and Challenges*; Sheehan, D., Ed.; American Institute of Physics: San Diego, CA, USA; 1411, 82–100.

[131] D. P. Sheehan (Ed.) (2011) *Second Law of Thermodynamics: Status and Challenges*, AIP Conf. Proc., 1411, 1. (doi:10.1063/1.3665227)

[132] Sheehan, D.P. and Wright, J.H. (2011) "Experimental Measurements of Electric Fields in Diodic Air Gaps: Toward a Second Law Challenge", in *Second Law of Thermodynamics: Status and Challenges*, AIP Conf. Proc., 1411, 147. (doi:10.1063/1.3665236)

[133] Versteegh, M. A. M., Dieks, D. (2011) "The Gibbs paradox and the distinguishability of identical particles", *American Journal of Physics*, 79(7), 741–746.

[134] D'Abramo, G. (2012) "A note on Solid-State Maxwell Demon", *Foundations of Physics*, Vol. 42, No. 3, 369-376.

[135] D'Abramo, G. (2012) "The peculiar status of the second law of thermodynamics and the quest for its violation", *Studies in History and Philosophy of Modern Physics*, 43, 226–235.

[136] Fu, X., Fu, Z. (2012) "Realization of Maxwell's hypothesis", http://arXiv.org/abs/physics/0311104 [physics.gen-ph].

[137] D'Abramo, G. (2013) "The Demon in a Vacuum Tube", *Entropy*, 15(5), 1916–1928.

[138] D'Abramo, G. (2014) ""Closed-loop" analysis of a thermocharged capacitor", arXiv:1407.6627 [physics.gen-ph].

[139] Sheehan, D.P., Mallin, D.J., Garamella, J.T, Sheehan, W.F. (2014) "Experimental Test of a Thermodynamic Paradox", *Foundations of Physics*, 44(3), 235–247.

[140] Fu, X., Fu, Z. (2014) "A Heat-Electric Conversion Caused by Two Different Work Functions", personal communication.

[141] Farley, F.J.M. (2015) "Muons, gravity and time", `http://arxiv.org/abs/1508.02339` [physics.gen-ph].

[142] *Limits to the Second Law of Thermodynamics*. (2016). Conference organizer: D. P. Sheehan, University of San Diego, San Diego, California.

Index

Synopsis This book brings together the chief results of the research work carried out by the author on the second law of thermodynamics and the theory of special relativity since 2008. The first six chapters are devoted to the research on the epistemological status of the second law of thermodynamics and the connection between thermionic/photoelectric phenomena and the second law: evidence is provided that thermionic emission could, in principle, violate the second law. More precisely, the photoelectric emission induced by the high-frequency tail of black-body radiation at room temperature (heat) can be harnessed to charge a capacitor and provide readily usable energy from a single heat reservoir. Chapter 7 contains some reflections on special relativity. It is the most speculative part of the book and it has not been published elsewhere. Two thought experiments on time dilation in the framework of special relativity are presented. The main contention in this part of the book is that if both postulates of special relativity are assumed to hold concurrently, then the prediction of asymmetric ageing made by Einstein in his 1905 relativity paper appears to be in fact incompatible with them and the fact that time dilation (which is intimately related to "asymmetric ageing") seems to have been experimentally confirmed provides, paradoxically, a refutation rather than a confirmation of the theory of special relativity, at least as interpreted today. A critical assessment of Purcell's basic explanation of magnetic forces, which basically relies on special relativity, is also given at the end of the book.

ABOUT THE AUTHOR. Germano D'Abramo worked as a researcher in various research institutes and companies in Rome between 1998–2014 on near-Earth asteroid population modeling and deflection. His research activity and interests also include algorithmic complexity, the second law of thermodynamics, probability, and the history and philosophy of physics. He is currently a high-school physics teacher. He is the author of the book *The Impact Clan* (CreateSpace Independent Publishing Platform; November 7, 2015).

www.ingramcontent.com/pod-product-compliance
Lightning Source LLC
Chambersburg PA
CBHW041241200526
45159CB00028B/9